AQA
A Level Maths

Year 1 + Year 2

Authors
John Rayneau,
Katie Wood

STATISTICS STUDENT WORKBOOK

Powered by **MyMaths**.co.uk

OXFORD
UNIVERSITY PRESS

Great Clarendon Street, Oxford, OX2 6DP, United Kingdom

Oxford University Press is a department of the University of Oxford.

It furthers the University's objective of excellence in research, scholarship, and education by publishing worldwide. Oxford is a registered trade mark of Oxford University Press in the UK and in certain other countries

First published in 2018

British Library Cataloguing in Publication Data

Data available

978-0-19-841308-0

3 5 7 9 10 8 6 4 2

Paper used in the production of this book is a natural, recyclable product made from wood grown in sustainable forests.

The manufacturing process conforms to the environmental regulations of the country of origin.

Printed and bound by CPI Group (UK) Ltd, Croydon, CR0 4YY

Acknowledgements

Authors
John Rayneau, Katie Wood

Editorial team
Dom Holdsworth, Ian Knowles, Matteo Orsini Jones

With thanks to Katherine Bird, Amy Ekins-Coward

Although we have made every effort to trace and contact all copyright holders before publication this has not been possible in all cases. If notified, the publisher will rectify any errors or omissions at the earliest opportunity.

Contents

About this book

This book provides additional support for the Statistics sections of the AQA AS Level and A Level Maths exams (7356 and 7357).

The Large data set

The AS and A Level examinations will assume that you are familiar with a Large data set (LDS). In the exam, some questions will be based on the LDS and may include extracts from it. It is the exam board's intention that you should be taught using the LDS, as this will give you a material advantage in the exam.

The introductory chapter to this book contains background info, interesting facts, and further explanations to all aspects of the data. This should provide you a solid foundation for understanding, interpreting and using the LDS throughout your examinations.

Corresponding Student Book Chapters

Inside this book, there is a chapter and section that corresponds to each Statistics chapter and section in OUP's AS and A Level Student Books that form part of this series (ISBNs 978 019 841295 3; 978 019 841296 0; 978 019 841294 6).

Inside each section you will find a recap of the content covered in the Student Book chapter, further worked examples, exam tips, and exam-style practice questions – so that you can turn up to your exams as well-prepared as possible. Each section, just like the Student Books, also contains MyMaths codes for further online practice.

Answers

The back of this book contains short answers to all questions.
Fully worked solutions are also available, password-protected for teacher access only, online.

   https://global.oup.com/education/content/secondary/series/aqa-alevel-maths/aqaalevelmaths-answers

Formulae

At the back of this book, you will find the Statistics formulae that you will be provided in an exam, as well as those that you will need to learn.

Calculators

All papers are calculator papers. You must therefore make sure that you have a calculator and that you know how to use it for all topics covered. The rules on which calculators are allowed can be found in the Joint Council for General Qualifications document 'Instructions for conducting examinations' (ICE).

The Large data set (LDS)

The examinations will assume that you're familiar with a Large data set (LDS). In the exam, some questions will be based on the LDS and may include extracts from it; you don't need your own copy of the data, and there's no need to memorise it. It is Ofqual's and AQA's intention that you should be taught using the LDS, as this will give you a material advantage in the exam. If you have background knowledge of the LDS, you'll be able to fully understand questions based on the LDS, without detailed explanations of the terminology, and you'll be able to put your answers into context.

The AQA LDS is based on extracts of the data used in a DEFRA (Department for Environment, Food and Rural Affairs) report published in 2015, 'Family Food'. The data set 'Purchased quantities of household food & drink by Government Office, Region and Country' gives the average quantities of certain foodstuffs purchased per person per week during the period 2001 to 2014, broken down by government administrative region. Further information can be found at

https://www.gov.uk/government/collections/family-food-statistics

An Excel spreadsheet containing the LDS is available from the AQA website.

http://www.aqa.org.uk/subjects/mathematics/as-and-a-level/
mathematics-7357/assessment-resources

This is real, uncleansed data and as such it could contain errors.

AQA plans to change the LDS during the lifetime of the specification, and so we advise always checking the AQA website for the latest version.

The survey

A national food survey has been conducted every year since 1940 and its details have changed several times. For example, prior to 2006, the data was reported for financial years, whilst from 2006 onwards it has been reported for calendar years.

The data used in the LDS is taken from the 'Family Food Module' of the 'Living Costs and Food Survey'. The overall survey is run by the Office for National statistics (ONS), with the Family Food Module being sponsored by DEFRA, who have a coordinating role on food policy across government. It is based on ~6000 private households, whose occupants volunteer to keep detailed diaries recording food purchases for a series of two-week periods. The periods were spread throughout the year in order to avoid seasonal biases. The sample size means that it is possible to stratify it by region and demographic characteristics: age, income group, occupation, ethnicity, urban/rural, etc.

The sample

Using the Post Office's list of addresses, a multistage, stratified sampling technique with clustering was used to select an initial 11 484 addresses in Great Britain. Interviewers then contacted each address, at least four times throughout the day, to invite the household to participate. Only 10 349 addresses were eligible households (not empty or businesses) and of these, 5144 agreed to participate, with 4760 households fully participating. The survey had a response rate of 46%, which is typical of major government surveys.

Each member of the household above 16 years of age was asked to keep a diary of food expenditure for specific two week periods. Children aged 7 to 16 were given a simplified diary. The data is backed up with actual till results to improve accuracy. The focus on expenditure, rather than consumption, is designed to avoid the under-reporting that occurs when people record consumption. In 2014, this resulted in 121 250 individual diaries.

Sources of potential errors include non-response bias, sample variability and incorrect reporting of certain items.

Bias and sample weighting

Since it is known that households who decline to respond to the survey differ from those who participate, sample weighting was used to compensate for their absence. Note, in the examinations, You will not be required to use weighting. The idea is that each respondent is given a weighting so that they also represent non-respondents who share similar survey characteristics. For example, if 10 households are identified as having the same characteristics but only 6 participate in the survey, then participants' results are given a weight of 1.6667 (= 10 ÷ 6). If there are N measurements, x_i, with weights w_i, then the average value of x, $\mu(x)$, is given by

$$\mu(x) = \frac{\sum_i w_i x_i}{\sum_i w_i} \xrightarrow{\text{Equal weights, } w_i = 1} \frac{\sum_i x_i}{N}$$

In practice, the characteristics of households who did and did not respond were determined by matching addresses to those in the preceding national census and using census data to identify the household characteristics and match respondents with similar non-respondents. Weights were used to correct for under-representation of people in the sample from government office region, metropolitan area, household composition, marital status, etc.

A second stage of weighting is used to compensate for the fact that only private households are included in the survey, which excludes people living in bed-and-breakfast accommodation, residential homes, hostels, etc. Four times a year the weights are adjusted so that weighted totals give the correct population totals, based on projections from the 2001 census, for gender, age group and regional populations.

The size of the survey means that it can be stratified using the nine government administrative regions. These are historic division of England, still used for statistical reports.

Region	Population (2011 census)	Median income (2014)
North East	2 596 886	£24 876
North West	7 052 177	£25 229
Yorkshire and the Humber	5 283 733	£24 999
East Midlands	4 533 222	£25 027
West Midlands	5 601 847	£24 920
East of England	5 846 965	£26 830
London	8 173 941	£35 069
South East	8 635 750	£28 629
South West	5 288 935	£25 571

For reference, in 2015 the average household spent £42.43 on food and drink per person per week. Eating in accounted for £29.24 and eating out £13.18. This represents ~11% of all household expenditure, rising to ~16% for people in the lowest 20% by equivalised household income.

As well as median income, London is exceptional in other ways. For example, the percentage of its population identifying as white is 59.8% compared to 86.8% in Britain as a whole. The percentage of the population born abroad is 36.7% compared to 18.3% in Britain as a whole.

More details of the survey's methodology can be found here:

https://www.ons.gov.uk/peoplepopulationandcommunity/personalandhouseholdfinances/incomeandwealth/compendium/familyspending/2015/methodology

Accuracy

Incorrect and inaccurate reporting is an issue. The diary sheets allow the recording of both the amount spent and quantity purchased for each individual item. Attaching till receipts to the diary sheets helps guard against errors but isn't always possible. During household visits, interviewers attempt to identify and correct reporting errors. However, they must balance this against the need to maintain the goodwill of participants and hence maintain high response rates: some poor data is better than no data!

Code	Description	Units	2001–02
22	Natural cheese	g	92
2201	Hard cheese - cheddar type	g	54
2202	Hard cheese - other uk or foreign equivalent	g	18
2203	Hard cheese - edam or other foreign	g	5
2205	Cottage cheese	g	9
2206	Soft natural cheese	g	7
2301	Processed cheese	g	14

If a quantity is missing, then a validation team member can add a 'proxy quantity' by searching a supermarket web site using the item description and amount spent on it. The food is classified using a hierarchical system of ~500 codes. If there is insufficient detail in an item's description, then a default code may be used based on the most commonly occurring product within a category. For example, 'cheese' would be coded as 'natural cheese, hard cheese – cheddar type' as this is the most popular type of cheese.

Some food is purchased 'when eating out', typically in the form of a meal. In this situation the participant only records the meal components and then a set of standard portion sizes are used to assign quantities.

The survey makes a distinction between food consumed at home and eating out expenditure. By this criterion school milk counts as eating out whereas take away food brought home counts as eating in.

Data quality and trend measures

Columns U – X in the spreadsheet give information on the reliability of the data and indicate whether there is any significant change in the data over the four year period, 2011–2014. This period is chosen as it is considered long enough to show any trend whilst still remaining current at the time of publication.

The definition of these quality and trend measures is technical and largely beyond what is required for the examinations. It is included here for completeness.

Relative Standard Error (RSE) indicator

Suppose there are N households, that household i contains n_i people and purchases an amount x_i of a good in one week. Then the average amount of x purchased per person, $\mu(x)$, per week and the standard error on this average, $\sigma(x)$ are given by

RSE indicator[a]	% change since 2011	sig[b]	trend since 2011[c]
✓✓✓	−6	yes	
✓✓	−0		
	+17		
✓✓	−1		
✗	−97	yes	↘
✗	+129		↗
✓✓	−16	yes	↘
✓	−41	yes	↘
✓	−7		
✓✓	−6		

Note all figures in the data set are stored too many decimal places but shown rounded to the nearest integer. The '–0' entry is really −0.484599647376746

$$\mu(x) = \frac{\sum_i w_i x_i}{\sum_i w_i n_i} \qquad \sigma(x) = \mu(x)\sqrt{\frac{N}{N-1}\sum_i w_i^2\left(\frac{x_i}{\sum_i w_i x_i} - \frac{n_i}{\sum_i w_i n_i}\right)^2}$$

The relative standard error is defined as the percentage error on the average.

$$\text{RSE} = \frac{\sigma(x)}{\mu(x)} \times 100\%$$

The smaller the value of the RSE, the more reliable the data. The (un)reliability of the data is indicated by assigning a quality symbol: $0.0\% \le \text{RSE} < 2.5\%$, ✓✓✓; $2.5\% \le \text{RSE} < 5.0\%$, ✓✓; $5.0\% \le \text{RSE} < 10\%$, ✓; $1.0\% \le \text{RSE} < 20\%$, 'blank'; $20\% \le \text{RSE}$ ✗

Percentage change is calculated in terms of the average values of a variable in 2011 and 2014, μ_{11} and μ_{14}, as

$$\frac{\mu_{14} - \mu_{11}}{\mu_{11}} \times 100\%$$

The **significance** of this change is flagged 'yes' if the difference in average values is over twice the standard deviation of the difference measurement.

$$\frac{|\mu_{14} - \mu_{11}|}{\sqrt{\sigma_{14}^2 + \sigma_{11}^2}} > 2$$

Assuming a Normal distribution, a two-standard deviation fluctuation from zero only has only a 4.55% probability of occurring by chance.

Trend since 2011 is provided as a check of whether there is a short-term trend. The gradient, m, of the line of best fit for the last four data points is compared to the standard deviation of the gradient, $\sigma(m)$. If this is greater than two, $|m| / \sigma(m) > 2$, and so unlikely to be a fluctuation of a genuinely zero trend, then the change is flagged as significant and increasing or decreasing using the symbols '↗' and '↘' respectively.

The relationships between topics

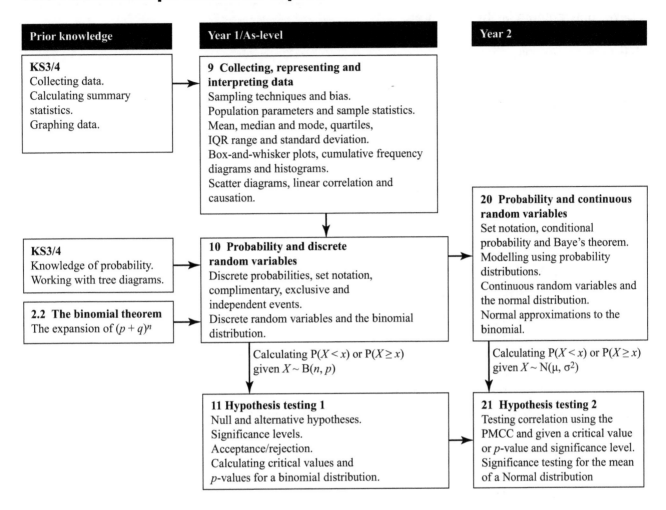

Prior knowledge	Year 1/As-level	Year 2
KS3/4 Collecting data. Calculating summary statistics. Graphing data.	**9 Collecting, representing and interpreting data** Sampling techniques and bias. Population parameters and sample statistics. Mean, median and mode, quartiles, IQR range and standard deviation. Box-and-whisker plots, cumulative frequency diagrams and histograms. Scatter diagrams, linear correlation and causation.	
KS3/4 Knowledge of probability. Working with tree diagrams. **2.2 The binomial theorem** The expansion of $(p + q)^n$	**10 Probability and discrete random variables** Discrete probabilities, set notation, complimentary, exclusive and independent events. Discrete random variables and the binomial distribution.	**20 Probability and continuous random variables** Set notation, conditional probability and Baye's theorem. Modelling using probability distributions. Continuous random variables and the normal distribution. Normal approximations to the binomial.
	Calculating $P(X < x)$ or $P(X \geq x)$ given $X \sim B(n, p)$	Calculating $P(X < x)$ or $P(X \geq x)$ given $X \sim N(\mu, \sigma^2)$
	11 Hypothesis testing 1 Null and alternative hypotheses. Significance levels. Acceptance/rejection. Calculating critical values and p-values for a binomial distribution.	**21 Hypothesis testing 2** Testing correlation using the PMCC and given a critical value or p-value and significance level. Significance testing for the mean of a Normal distribution

Notation and language

A Level-only content is shown in grey.

Probability and Statistics

A, B, C, etc.	events
$A \cup B$	union of the events A and B
$A \cap B$	intersection of the events A and B
$P(A)$	probability of the event A
A'	complement of the event A
$P(A \mid B)$	probability of the event A conditional on the event B
X, Y, R, etc.	random variables
x, y, r, etc.	values of the random variables X, Y, R etc.
$x_1, x_2, \dots$	observations
$f_1, f_2, \dots$	frequencies with which the observations $x_1, x_2, \dots$ occur
$p(x)$, $P(X=x)$	probability function of the discrete random variable X
$p_1, p_2, \dots$	probabilities of the values $x_1, x_2, \dots$ of the discrete random variable X
$E(X)$	expectation of the random variable X
$Var(X)$	variance of the random variable X
$\sim$	has the distribution
$B(n, p)$	binomial distribution with parameters n and p, where n is the number of trials and p is the probability of success in a trial
q	$q = 1 - p$ for binomial distribution
$N(\mu, \sigma^2)$	Normal distribution with mean μ and variance σ^2
$Z \sim N(0,1)$	standard Normal distribution
ϕ	probability density function of the standardised Normal variable with distribution $N(0, 1)$
Φ	corresponding cumulative distribution function
μ	population mean
σ^2	population variance
σ	population standard deviation
$\overline{x}$	sample mean
s^2	sample variance
s	sample standard deviation
H_0	null hypothesis
H_1	alternative hypothesis
r	product moment correlation coefficient for a sample
ρ	product moment correlation coefficient for a population

Recap

- A **sample** is a selection of people or things taken from the whole **population**.
- **Parameters** describe the whole population. **Statistics** derived from the sample are used to estimate the parameters of the population.

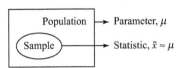

| All the people in the UK is a population. | The mean age of all the people (μ) is a parameter. |
| A selection of people from the UK is a sample. | The mean age of the people in the sample ($\bar{x}$) is a statistic. |

Choosing a different sample may lead to a different sample mean.

There are many different approaches to finding a sample.

If you have a list of the whole population, then one or more of these methods can be used.

Simple random sampling	**Systematic sampling**	**Stratified sampling**
The sample is chosen completely randomly: every member of the population is equally likely to be chosen using this method.	A rule is used to select the sample, such as selecting every nth item. It is *not* fully random so there is a potential for bias.	The population is split into categories and the proportion of each category in the sample is the same as the proportion in the whole population.

Sometimes, there may not be a list of the members of the population, in which case one (or a combination) of these methods can be used.

Opportunity sampling	**Quota sampling**	**Cluster sampling**
You select from an accessible part of the population: there is a risk of bias.	The size of the sample for each category is chosen in advance. Use with opportunity sampling to reduce bias.	The population is divided into clusters with similar characteristics and then certain clusters are chosen by simple random sampling.

When deciding on a sampling method, cost, difficulty and time constraints need to be considered. However, it is important to try to avoid **bias**. A biased sample will not be representative of the population and can lead to inaccurate estimations of the population parameters.

You may be asked to explain why sampling methods are biased.

Sampling method	Reason for bias
Sampling people outside a library when investigating the reading habits of people in a town.	People visiting a library may have different reading habits to the population as a whole.
Randomly choosing 20 men and 20 women in a gym to complete a survey about their exercising habits when 80% of the gym members are male.	The sample is not representative of the population, which is problematic as men and women may have different exercising habits.

Example 1

A crop of runner beans comes from three different varieties of seed. There are 15 plants of type A, 45 of type B and 30 of type C. A farmer wishes to estimate the mean yield of the crop. She proposes to count the number of beans produced by a sample of three plants, choosing one from each type of seed.

a   Explain any disadvantages of this proposal.

b   Work out how many plants of type C should be selected for a stratified sample of size 12

a   The sample will not be representative of the population; the different varieties of plant may perform differently. The sample size is too small to make an accurate estimation of the population mean.

There are 90 plants in total and 30 of type C.

b $\dfrac{30}{90} \times 12 = 4$ of type C.

Notice that the proportion of the sample that is plant C is the same as the proportion of plant C in the population.

Example 2

A researcher is investigating which breeds of dog are the most popular in her city. She splits the city into areas of roughly equal size, then chooses two areas at random and looks at all dogs registered within those two areas.

a   What is this sampling method called?

b   Explain an advantage and disadvantage of this sampling method.

a   Cluster sampling.

The clusters are the areas of the city.

b   An advantage is that it is easier to collect the sample in just two areas rather than all over the city.

A disadvantage is that the breeds of dog in those two areas may not be representative of the city as a whole.

Exam tips

- Make sure you explain your answers thoroughly. For example, don't just say 'opportunity sampling may introduce bias', but explain *why* or *how*.
- When describing a sampling method, give full details and be as precise as you can.
- Always give your answers in the context of the question.
- Remember that you may need to use your knowledge of the Large data set, so make sure that you are familiar with it.

Exam practice questions

1   Brock wants to investigate the attitude of his year group to the consumption of meat and fish. In order to obtain a sample, he writes the names of all of the boys on separate slips of paper and places these in a box. He repeats this for the girls using a separate box. He then takes a simple random sample of names from each box where the number of names sampled from each box is proportional to the number of names in that box.

   a   What is the population from which Brock is sampling?   [1]

   b   Name the type of sampling he is using.   [1]

   There are 120 students in the year group, of which 80 are girls.

   c   If Brock wants to take a 10% sample of the population, how many boys and how many girls should he have in his sample? You should show details of any calculations.   [2]

2   A machine cooks jam tarts by placing uncooked tarts on a conveyor belt and running them under heating elements for exactly 12 minutes. Samples are taken to ensure high quality.

   a   If the temperature of the heating elements and all other conditions in the factory remain unchanged throughout the day, explain why sampling over a single, short period of time is the best method.   [2]

   b   If the temperature of the heating elements rises slightly throughout the day, why might systematic random sampling be more suitable?   [2]

3 In order to investigate meat consumption in different regions, researchers for the survey Family Food use the Postcode Address File, the Post Office's list of UK addresses, to divide all addresses into different regions. A simple random sample of households is taken from each region and all adult members of each sampled household are questioned.

a Name two methods of sampling illustrated here apart from the simple random sampling within different regions, and explain why each might be used. [3]

b Bias could be introduced if the households were not initially divided into regions. Explain why. [1]

4 A researcher wishes to investigate public opinion on the repealing of the 2004 Hunting Act. He goes to a country pub one Sunday lunchtime and asks the views of the first 20 people to arrive.

a Give two possible sources of bias in this sampling procedure. [2]

b State how each source of bias may be eliminated. [2]

5 Describe a systematic sampling procedure for obtaining a random sample of 10% of the pupils in a school so that the sample reflects the population ratio of boys to girls and the full age range of pupils is sampled. [3]

6 Research is being carried out into fruit sales in the UK between the years 2010 and 2015.

a Why is it necessary to compare sales during the same months for each year? [2]

6 b Explain why it is particularly important to stratify the whole population of households of the UK by country if the total sample size is small. [2]

_____

_____

7 As statistical consultant to Northern Light Dairy, you have been asked to investigate regional variation and the effect of income on the purchasing of dairy products in the north of England and the Midlands.

You obtain the following data, which gives the populations, in thousands to the nearest thousand, of five of the nine English regions with the proportion of income support claimants in each region.

Region	North East	North West	Yorkshire/ Humber	East Midlands	West Midlands
Population (thousands)	2597	7052	5284	4533	5602
% claiming income support	3.0	5.3	5.3	3.5	5.1

You wish to take a stratified random sample of total size 1000 from the five regions.

a Show that the number of non-claimants and claimants from the North East in the sample should be 3 and 100, respectively. [4]

_____

_____

_____

_____

_____

_____

The total amount of whole milk bought per week for all members of the sample from the North East was 38.643 litres.

b Estimate the total amount of milk that the dairy can expect to sell per week to residents from the North East. [2]

_____

_____

_____

_____

Recap

- **Mode**, **median** and **mean** are measures of **central tendency** or **average**.
- **Range**, **interquartile range**, **variance** and **standard deviation** are measures of **spread**.
- You should be able to identify **outliers** and explain their effect.

Mode
The most common value or the group with the highest frequency.

Median
The middle value when the data is listed in size order.

Mean
The sum of the values divided by how many there are: $\mu = \dfrac{\sum x}{n}$

The number of portions of fruit or vegetables eaten per person per day in a sample is: 9, 2, 1, 3, 2, 1, 2, 2, 3, 3, 5, 5

- The mode is 2 as this is the most commonly occurring value.
- To find the median, you must put the numbers in *size order*: 1, 1, 2, 2, 2, 2, 3, 3, 3, 5, 5, 9
- Since there are an even number of values the median is the average of the two in the middle, so $\dfrac{2+3}{2} = 2.5$ portions.
- To find the mean, add up the values and divide by 12: $\dfrac{38}{12} = 3.17$ portions.

- The range of the number of portions of fruit or vegetables is: $9 - 1 = 8$
- To find the quartiles, use the ordered list of the data: 1, 1, 2, 2, 2, 2, 3, 3, 3, 5, 5, 9

  $\dfrac{12+1}{4} = 3.25$ so the lower quartile is the average of the 3rd and 4th values: $\dfrac{2+2}{2} = 2$ portions.

  $\dfrac{3(12+1)}{4} = 9.75$ so the lower quartile is the average of the 9th and 10th values: $\dfrac{3+5}{2} = 4$ portions.
- Therefore the interquartile range is: $\text{IQR} = 4 - 2 = 2$
- To find the variance, square each of the values and add them up: $9 + 2 + 1 + \ldots$, then the variance is given by

  $\sigma^2 = \dfrac{176}{12} - \left(\dfrac{38}{12}\right)^2 = 4.64$
- The standard deviation is the square-root of the variance, so $\sigma = \sqrt{4.64} = 2.15$
- The formula for standard deviation will be given to you in the examination as:

  standard deviation $= \sqrt{\dfrac{\sum x^2}{n} - \bar{x}^2}$ where $\bar{x}$ is the mean.

Range
The difference between the smallest and largest values.

Interquartile range
The difference between the upper and lower quartiles: $\text{IQR} = Q_3 - Q_1$

Variance
A measure of the spread of the data: $\sigma^2 = \dfrac{\sum x^2}{n} - \left(\dfrac{\sum x}{n}\right)^2$

Standard deviation
The square-root of the variance: $\sigma = \sqrt{\dfrac{\sum x^2}{n} - \mu^2}$

- You need to be able to explain which average is the best to use in a particular situation.

In the example above, the median or mode is preferable as they are not affected by the value of 9 which is an **outlier**. You will be given a rule to use to decide if a value is an outlier.

Example 1

The table gives the mark of students in a statistics exam to the nearest 1%

Mark (%)	Number of students
0–19	1
20–39	23
40–59	27
60–79	35
90–100	40

a   Estimate the mean and standard deviation.

b   Estimate the median.

A student was incorrectly recorded as having scored 0% when in fact they scored 100%

c   Without further calculation, explain the effect that rectifying this error will have on the mean and the standard deviation.

a $\sum x = (9.5 \times 1) + (29.5 \times 23) + (49.5 \times 27) + (69.5 \times 35)$
$+ (90 \times 40) = 8057$

n is the total number of students, which is 126

So the mean is: $\mu = \dfrac{8057}{126} = 63.9\%$

$\sum x^2 = (9.5^2 \times 1) + (29.5^2 \times 23) + (49.5^2 \times 27) + (69.5^2 \times 35)$
$+ (90^2 \times 40) = 579321.5$

So the standard deviation is $\sigma = \sqrt{\dfrac{579321.5}{126} - \left(\dfrac{8057}{126}\right)^2} = 22.6$

> Since the data is grouped, use the midpoints as estimates for x

> Remember: you should be able to do these calculations using the STAT mode on your calculator.

b $\dfrac{126}{2} = 63$ so you need to estimate the 63rd value.
This lies in the 60–79 group and is the 12th value of the 35 in the group.

So the median is estimated as: $59.5 + \dfrac{12}{35} \times 20 = 66.4\%$

> Since we are estimating the median, there is no need to use $n + 1$

c   The mean will be higher and the standard deviation will be smaller.

Exam tips

- Remember that the range is a single number. In everyday language you might say 'the range is 6 to 15', but mathematically the range is $15 - 6 = 9$
- Check whether you are asked for the variance or standard deviation and remember that the standard deviation is the square root of the variance.
- When calculating the mean of grouped data, be careful to divide by the total number of items of data and not by the number of groups.
- Only round your answers at the end of a question so you don't make mistakes due to premature rounding.
- Give answers to at least 3 significant figures unless advised otherwise.
- When comparing two sets of data, use statistical vocabulary such as mean, standard deviation, median etc. and state the actual values you are comparing.

Exam practice questions

1   For the following sets of data, find the median and the interquartile range.

a   12.5 13.5 11.8 20.2 17.4 18.7 15.6 14.9 12.0 13.0   [4]

b

x	0	1	2	3	4
f	5	9	12	8	1

[3]

2   Find the mean and variance of the following data sets.   [2]

a   12 16 17 18 11 12 14 9 13 15 20 8

b

x	−2	−1	0	1	2
f	2	7	9	6	1

[2]

3   The mean vitamin C intake for 34 children in the North East of England is 66 mg per day. The equivalent statistic for 18 children in the South East is 71 mg. Find the mean intake for all 52 children.   [3]

4 The quantities of soft drinks, including milk, bought in millilitres by 14 teenagers in Scotland over a period of one week are shown.

$$246\ 265\ 231\ 163\ 310\ 267\ 241\ 219\ 233\ 251\ 276\ 284\ 309\ 231$$

a Find the mean, variance and standard deviation for this data. [3]

An outlier is defined as any observation outside the interval (mean $\pm\ 2 \times$ standard deviation).

b Calculate this interval for the data and state whether any of the observations should be classified as outliers. [3]

5 The percentage daily maximum relative humidity for 31 days at a weather station during the summer of 2015 is given in the table.

x	$75 \leq x < 80$	$80 \leq x < 85$	$85 \leq x < 90$	$90 \leq x < 95$	$95 \leq x < 100$
f	1	6	7	8	9

a Find the mean and variance of the data, giving your answer to 2 dp. [2]

b A day is chosen at random. Find the probability that the maximum relative humidity on this day is within one standard deviation of the mean. [5]

6 Two samples of adults from two towns in south-west England are taken and, for each adult, the amount of butter bought over a period of 4 weeks is noted. The data gives the following statistics.

	Mean	Variance	Sample size
Town A	134.1	222.2	17
Town B	130.2	201.9	8

a Find Σx and Σx^2 for each of the samples. [5]

b Use your results to find the mean and variance of the combined sample of 25 values. [3]

c Show that the mean of the combined sample is NOT equal to the mean of the two mean values and explain why this is the case. [2]

d The **weighted mean** of two samples of sizes n_1 and n_2 and means $\bar{x}_1$ and $\bar{x}_2$ is given by $\bar{x}_w = \dfrac{n_1 \bar{x}_2 + n_2 \bar{x}_2}{n_1 + n_2}$. Show that the mean value found in part **b** is the weighted mean of the two samples. [2]

Recap

- **Continuous data** can take any value in a given range.
- You can represent continuous data using one of these graphs:
    - A **cumulative frequency graph** can be used to estimate the median and quartiles.
    - A **box-and-whisker-plot** shows the max. and min. values and the quartiles.
    - A **histogram** shows the distribution of the data and the area of each bar represents the frequency.

The cumulative frequency curve shows the ages of people at a party.

There are 40 people in total so the median is the 20th person–from the graph, you can estimate this person to be 17 years old.

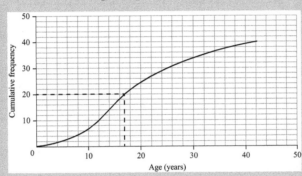

Use a similar method to see that the lower quartile (the 10th person) is 12 years old and the upper quartile (the 20th person) is 25 years old.

The same statistics can be summarised in a box and whisker diagram.

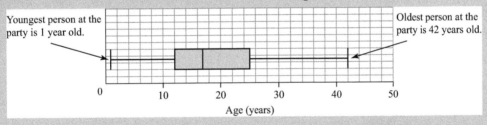

Youngest person at the party is 1 year old.

Oldest person at the party is 42 years old.

When interpreting a histogram, you must remember that the y-axis gives the **frequency density** and that:

frequency $\propto$ area

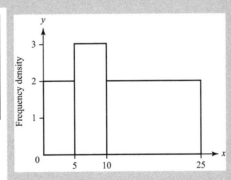

The 0–5 bar has an area of $5 \times 2 = 10$

The 5–10 bar has an area of $5 \times 3 = 15$

The frequencies could be equal to or proportional to the areas.

Example 1

The table shows the amount of cream purchased per person per week for a sample of households in the North East of the UK.

A histogram is drawn and the $5 \le m < 10$ group is 2 cm wide and 4 cm tall.

Calculate the dimensions of the bar representing the $10 \le m < 20$ group.

Cream purchased per person, c (ml)	Number of households
$0 \le m < 1$	16
$1 \le m < 5$	10
$5 \le m < 10$	24
$10 \le m < 20$	6

The $5 \le m < 10$ group has a class width of 5 ml and a width of 2 cm. Therefore, since the $10 \le m < 20$ group has a class width of 10 ml, the width of the bar will be 4 cm.

The area of the $5 \le m < 10$ group is $2 \times 4 = 8\,cm^2$.

From the table we can see that this group has a frequency of 24, so the relationship between frequency and area is:

Frequency = 3 × area.

The $10 \le m < 20$ group has a frequency of 6 so the area must be $6 \div 3 = 2\,cm^2$.

Therefore the height of the bar is $2 \div 4 = 0.5\,cm$

Width = 4 cm, height = 0.5 cm

- First work out the width of the bar.
- Calculate the area of the group given.
- Find the relationship between frequency and area for this particular histogram.
- Since height = area ÷ base

Example 2

Two maths classes sat a test and their results are summarised in this diagram.

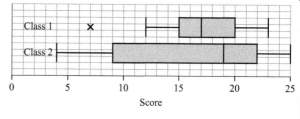

a Briefly compare the performance of the two classes.

b Explain what is shown by the cross.

c The pass score for the test was 15. Which class had the higher pass rate?

a Class 2 scored higher on average as they had a higher median (19) than class 1 (17). However, the results for class 2 were much more varied with an interquartile range of 13 compared to 5 for class 1

b The cross represents an outlier.

c Class 1, since 75% of students achieved the pass mark. In class 2, this figure was between 50% and 75%.

- Comment on average and spread, giving numbers to back up your statements.
- You know this because the lower quartile is the value that 75% of people exceed.

Exam tips

- Always give your answers in the context of the question.
- When drawing histograms, remember that the values on the y-axis are frequency densities and it is the *area* that represents the frequency, not the height.
- Remember that outliers are not included in the whiskers on box-and-whisker plots, but are shown with a cross.
- Be careful when reading numbers from a cumulative frequency graph. For example, using the graph above, the number of people at the party over age 30 is $40 - 34 = 6$
- Remember that you may need to use your knowledge of the large data set, so ensure that you are familiar with it.

Exam practice questions

1 The average volume of regular and low-calorie soft drinks bought by a sample of 40 UK residents in 2012, in millilitres per week, is shown in the following box-and-whisker plots.

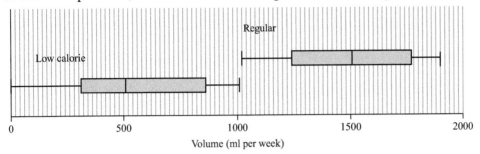

Volume (ml per week)

a Fill the table with the median, range and interquartile range for the volume of low-calorie soft drink purchased by members of the sample. Give all answers to the nearest 10 ml.   [3]

	Median	Range	Interquartile range
Regular			
Low calorie			

b Compare the distributions of the volumes of low-calorie and regular soft drinks purchased.   [2]

_____

_____

_____

2 34 households were asked about the amount of meat they consumed. The results, g, in grams per person per week, are shown in the following histogram. (The bar for $40 \leq g < 50$ is missing.)

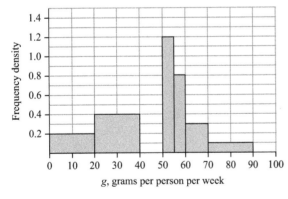

g, grams per person per week

a Find the frequencies for each interval and complete the graph by drawing the bar for $40 \leq g < 50$ [4]

_____

_____

_____

_____

Interval, g	$0 \leq g < 20$	$20 \leq g < 40$	$40 \leq g < 50$	$50 \leq g < 55$	$55 \leq g < 60$	$60 \leq g < 70$	$70 \leq g \leq 90$
Frequency							

2 b Calculate the median amount of meat consumed per person.   [2]

_____

_____

_____

One household is chosen at random.

c Find the probability that the expenditure for this household is above £50   [1]

3 The daily total number of hours of sunshine at a certain location over 30 days in the summer of 2015 are shown in the following cumulative frequency graph.

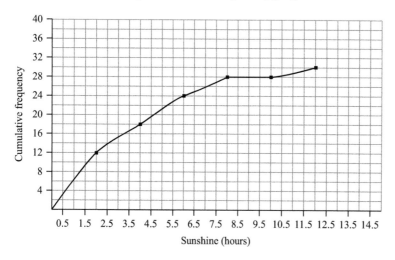

a Use the cumulative frequency curve to find

  i The median number of hours of sunshine,   [1]

  ii The interquartile range for the data,   [2]

_____

_____

_____

  iii The number of days for which the number of hours of sunshine was greater than 6   [1]

_____

_____

3 b On how many days is the number of hours of sunshine more than one interquartile range away from the median?   [3]

_____

_____

_____

4 The blood sugar levels of 23 patients with type 1 diabetes were recorded before a meal and 90 minutes after a meal. These cumulative frequency curves show the results.

The researchers had lost the original tabulated data.

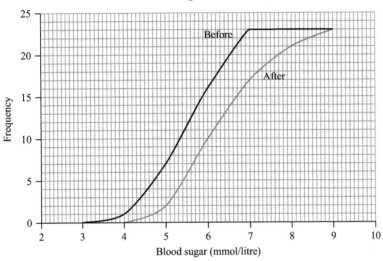

a Reconstruct the original data and complete the following table.   [2]

Blood sugar, b (mmol/litre)	$3 \leq b < 4$	$4 \leq b < 5$	$5 \leq b < 6$	$6 \leq b < 7$	$7 \leq b < 8$	$8 \leq b < 9$
Before						
After						

b Write down the median and lower and upper quartiles for each of the sets of data.   [4]

_____

_____

_____

_____

c Compare the blood sugar levels for the two groups.   [2]

_____

_____

_____

Recap

- A **scatter diagram** shows the relationship between two variables. This can be
    - Moderate or strong positive correlation,
    - Moderate or strong negative correlation,
    - No correlation.
- The **correlation coefficient**, r, gives an indication of the strength of the correlation. It is always in the range $-1 \le r \le 1$, so, for example, $r = 1$ indicates perfect positive correlation and $r = 0$ indicates no correlation.
- A **regression line** is a line of best fit. The closer the points lie to the line the stronger the correlation.

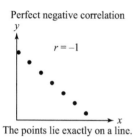

Perfect negative correlation

$r = -1$

The points lie exactly on a line.

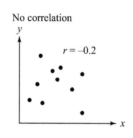

No correlation

$r = -0.2$

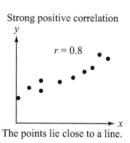

Strong positive correlation

$r = 0.8$

The points lie close to a line.

It is important to remember that correlation *does not imply causation.*

For example, a third variable (such as time) could be affecting both of the original variables.

A lack of correlation does not necessarily mean no relationship between the two variables–there could be a non-linear relationship between them.

In these cases, x is the known as the **independent** or **explanatory** variable and y is known as the **dependent** or the **response** variable.

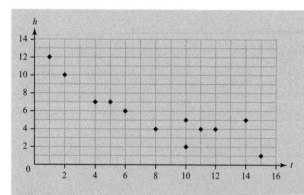

You can see that there is a strong negative correlation: the points lie close to a straight line.

So you could estimate the value of r to be between –0.9 and –0.8

A regression line has also been added.

The gradient of the line is negative since there is negative correlation.

Example 1

The amount of honey and jam purchased on average per person from the nine regions of England is shown in the scatter diagram.

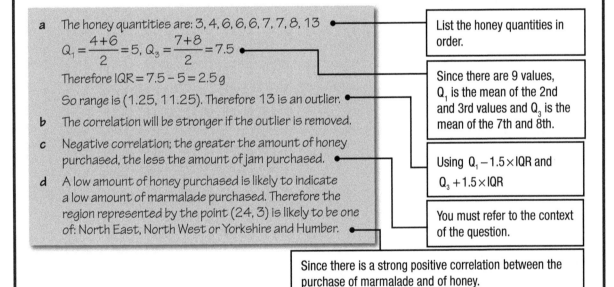

Honey purchased, h, (g) / Jam purchased, j, (g)

a Use the formula
$$\left(Q_1 - 1.5 \times \text{IQR} , \ Q_3 + 1.5 \times \text{IQR}\right)$$
to decide if there are any outliers in the amount of honey purchased.

b What effect will removing an outlier have on the strength of the correlation?

c Give an interpretation of the correlation if the outlier is removed.

The table summarises the amount of marmalade purchased on average per person in each of the nine regions of England.

The value of the correlation coefficient between the purchase of marmalade and of honey is 0.88

d Suggest which region could be represented by the point at $(24, 3)$

Region	Marmalade (g)
NE	4
NW	4
Y&H	4
EM	6
WM	5
E	6
L	6
SE	7
SW	10

a The honey quantities are: 3, 4, 6, 6, 6, 7, 7, 8, 13

$Q_1 = \dfrac{4+6}{2} = 5$, $Q_3 = \dfrac{7+8}{2} = 7.5$

Therefore IQR = 7.5 – 5 = 2.5 g

So range is (1.25, 11.25). Therefore 13 is an outlier.

b The correlation will be stronger if the outlier is removed.

c Negative correlation; the greater the amount of honey purchased, the less the amount of jam purchased.

d A low amount of honey purchased is likely to indicate a low amount of marmalade purchased. Therefore the region represented by the point (24, 3) is likely to be one of: North East, North West or Yorkshire and Humber.

> List the honey quantities in order.

> Since there are 9 values, Q_1 is the mean of the 2nd and 3rd values and Q_3 is the mean of the 7th and 8th.

> Using $Q_1 - 1.5 \times \text{IQR}$ and $Q_3 + 1.5 \times \text{IQR}$

> You must refer to the context of the question.

> Since there is a strong positive correlation between the purchase of marmalade and of honey.

Exam tips

- Always give your answers using the context of the question. For example, using the example above, don't just say there is negative correlation–instead, explain that as the amount of honey purchased increases, the amount of jam purchased decreases.
- Remember that whilst correlation can mean causation, it may also be coincidental.
- Remember that you may need to use your knowledge of the Large data set, so ensure that you are familiar with it.

1 Match each of the following pairs of variables with the most likely description of their
type and strength of correlation. [4]

    A Average daily temperature and amount spent per day on heating fuel.
    B The price of 1kg of cheese and the length of its maturation.
    C Foot length and shoe size.
    D A store's piano sales and the outdoor temperature.

    1 Strong positive correlation   **2** Zero correlation
    3 Moderate positive correlation   **4** Moderate negative correlation

2 The following data set gives the average amount purchased (in grams) per person per month for five
types of cheese for the South East (x) and the South West (y) of England in 2014

x	72	7	9	6	22
y	80	7	7	4	30

    a Draw a scatter diagram and use it to describe the type and degree of correlation. [4]

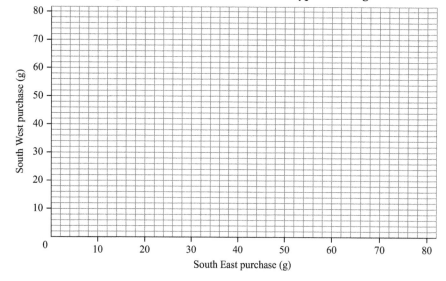

    b State the value of the correlation coefficient given that it is either –0.9, 0.8, 0.45 or 1.3 [1]

    c Explain the meaning of the term 'spurious correlation' and state whether the variables
    x and y exhibit this type of correlation. You should give a reason for your answer. [2]

3 The following table gives the reading age (R years) and results for a general knowledge test ($T\%$) for eight students.

R	8.4	13.1	12.0	9.2	6.4	8.4	10.1	12.2
T	47	73	61	58	41	52	58	50

a Draw a scatter diagram for this data with values of R on the horizontal axis. [2]

General knowledge result (%)

Reading age (years)

b Describe the correlation between reading age and test results. [1]

_____

c Explain why the scatter diagram should not be used to predict the test results of a student with a reading age of 18 years. [1]

_____

d On your scatter diagram, draw a line of best fit and use it to predict the test result for a student with a reading age of 11 years. [2]

_____

4 The diagram shows a spinner that can land on numbers 1–5 and on regions which are either plain (event A), lightly shaded (event B) or heavily shaded (event C).

An experiment consists of spinning the spinner 5 times and noting the number (1–5) and event (A, B or C) occurring for each spin.

N_1 is the number of 1s etc.

N_A is the number of As etc.

The experiment is performed 10 times.

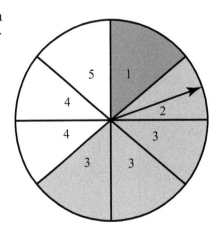

4 a State whether each of the following pairs of events show positive, negative or zero correlation. [4]

    i N_1 and N_C

    ii N_1 and N_5

    iii N_3 and N_B

    iv N_A and N_B

b Which of the above pairs of events has perfect correlation? Explain your answer. [2]

5 The following coordinates give a sample of size 4 from a bivariate population: $A(2.0, 2.7)$, $B(4.3, 4.2)$, $C(6.9, 3.3)$, $D(8.8, 2.2)$

    a Draw a scatter diagram for this data. [2]

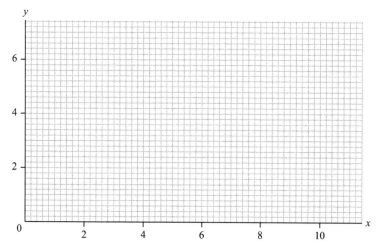

It turns out that the y-value of one of the points was erroneously recorded at half its correct value.

    b Identify the point and state its true y-value given that the data is

    i Strongly negatively correlated, [1]

    ii Moderately positively correlated. [1]

Recap

- The collection of all possible outcomes is called the **sample space**. The probabilities inside a sample space must add up to 1
- **Events** are groups of one or more outcomes.
- Two events are **mutually exclusive** if they cannot happen together.
- If A and B are mutually exclusive events, then
  - $P(A \cap B) = 0$
  - $P(A \cup B) = P(A) + P(B)$
- The probabilities of all the possible mutually exclusive events add up to 1
- A **probability distribution** is a table or function that gives the probability of all possible outcomes.

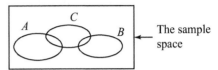

The sample space

A and B are mutually exclusive as there is no overlap between them.

$A \cap B$ is the intersection between A and B, i.e. they both happen.

$A \cup B$ is the union of A and B, i.e. at least one of them happens.

An example of a probability function is given in the table. The probability of event A occurring is 0.2. You could write this $P(A) = 0.2$

Since the probabilities must add up to 1,
$0.3 + 4k = 1 \Rightarrow k = 0.175$

Outcome	A	B	C	D
Probability of outcome	0.2	k	$3k$	0.1

Therefore the rest of the probabilities are:
$P(B) = 0.175$, $P(C) = 0.525$, $P(D) = 0.1$

- Two events are **independent** if they have no effect on each other.
- If events A and B are independent, then $P(A \cap B) = P(A) \times P(B)$

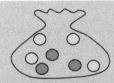 For example, if you randomly select a counter from a bag of different coloured counters and replace it before selecting the next counter, then you can assume that the colour of the first counter will not affect the colour of the second counter. Therefore, these events are independent.

If you are told that a bag contains a *large* number of counters, then you can assume independence even if the counters are not replaced.

For example, if in a large bag of counters, 40% of them are blue and 60% of them are red, then the probability of drawing one of each colour is given by

$P(\text{one of each}) = 2 \times P(\text{first counter blue}) \times P(\text{second counter red})$
$= 2 \times 0.4 \times 0.6 = 0.48$

You multiply by 2 because there are 2 ways of selecting one of each: blue followed by red or red followed by blue and both of these have the same probability.

Example 1

A sample of men and women are asked how many meals out they had yesterday. The results are summarised in the table.

Number of meals out	Men	Women	Total
> 1 meal out	4	3	7
Exactly 1 meal out	14	7	21
No meals out	10	11	21
Total	28	21	49

a Write down the probability that

   i A randomly selected person had no meals out that day,

   ii A randomly selected man had exactly 1 meal out.

Event A is defined as 'a randomly selected person is a man', event B is defined as 'a randomly selected person ate more than 1 meal out' and event C is defined as 'a randomly selected person ate no meals out'.

b Which pair of events are mutually exclusive?

c Are the events A and B independent? Use probabilities to justify your answer.

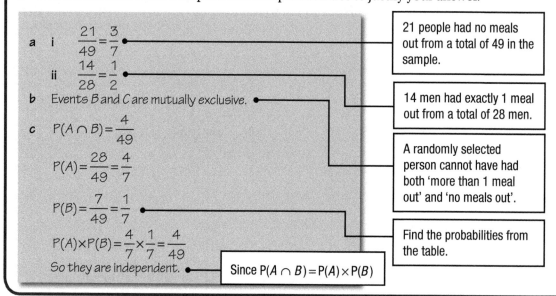

a i $\dfrac{21}{49} = \dfrac{3}{7}$ — 21 people had no meals out from a total of 49 in the sample.

  ii $\dfrac{14}{28} = \dfrac{1}{2}$ — 14 men had exactly 1 meal out from a total of 28 men.

b Events B and C are mutually exclusive. — A randomly selected person cannot have had both 'more than 1 meal out' and 'no meals out'.

c $P(A \cap B) = \dfrac{4}{49}$

$P(A) = \dfrac{28}{49} = \dfrac{4}{7}$ — Find the probabilities from the table.

$P(B) = \dfrac{7}{49} = \dfrac{1}{7}$

$P(A) \times P(B) = \dfrac{4}{7} \times \dfrac{1}{7} = \dfrac{4}{49}$

So they are independent. — Since $P(A \cap B) = P(A) \times P(B)$

Exam tips

- It is usually a good idea to draw a diagram to illustrate the situation. A Venn diagram or a tree diagram can often help you see what is going on.
- If you draw a Venn diagram, don't forget a box to define the universal set.
- Check carefully whether events are mutually exclusive or independent and don't mix up the definitions. Mutually exclusive events cannot happen together, and events are independent if the fact that one has happened does not affect the probability that the other will happen.
- Remember that probabilities must be between 0 and 1, so go back and check your working if you get an answer outside of this range!

1 **a** Describe each of the following pairs of events as either mutually exclusive or independent, giving
reasons for your answers.   [3]

    i   Two coins are thrown in order: Getting a head on the first and a tails on the second.

    ii One six-sided dice is thrown: Getting an even number and getting a 3

    iii One ball is chosen from a bag containing red and blue balls only: Getting a red ball and getting a
blue ball.

    b Which, if any, of the pairs of events are complementary?   [1]

2 I throw a coin once. If it lands heads up, I throw it again. If it lands tails up, I throw an ordinary
six-sided dice.

    a List all the outcomes that make up the sample space for this experiment.   [2]

    The sample space for this experiment contains n sample points.

    b Explain why it is not correct to assume that each outcome has a probability of $\dfrac{1}{n}$ [1]

3 Two ordinary six-sided dice are thrown and the difference in the scores (first score minus second score)
is calculated.

    a By drawing a two-way table to represent the sample space, find the probability that the value of the
difference is −1   [2]

3 b If the two dice are thrown twice, what is the probability of getting a difference of −1 followed by a difference of 3? [1]

4 a Two events, *A* and *B*, are associated with an experiment. Use set theory notation to indicate that the two events are independent. [1]

b A person's sodium intake is largely affected by the amount of dietary salt consumed. The table gives the proportions of adults in a large sample of UK residents with their sodium intake, in grams per day, broken down into three age groups.

Sodium intake		
Age (years)	**Below 2.9**	**2.9 or above**
Less than 50	0.34	*a*
50–70	0.09	0.14
Greater than 70	0.11	0.05

A person is chosen at random.

i Find the value of *a* [1]

ii What is the probability that the person has an sodium intake lower than 2.9 grams per day? [1]

iii What is the probability that the person is less than 50 years old? [1]

iv Use your answers to parts **ii** and **iii** to decide whether a person's sodium intake is independent of the person's age. You should show your working. [2]

5 A random variable, X, can take integer values 1–5. The probability distribution function of X is given by

$$P(X=x)=\begin{cases} a \text{ for } x=1,2,3,4 \\ 2a \text{ for } x=5 \\ 0 \text{ otherwise} \end{cases}$$

where a is a constant.

a Find a and $P(X<4)$ [2]

b Find the probability that, out of four X-values sampled, exactly two are greater than or equal to 4. State any assumptions you make. [2]

6 In 2014, the median amount of cheese purchased in the West Midlands was 102 grams per person per week. Find the probability that, when three people are chosen from this area,

a The first two buy more than 102 g and the third buys less than 102 g, [2]

b Two buy more than 102 g and one buys less than 102 g. [2]

Recap

- A **random variable** can be discrete or continuous.
- A discrete random variable has a **binomial distribution** if these conditions are met:
  - There are only two possible outcomes (e.g. success/failure).
  - There are a fixed numbers of trials (n).
  - The trials are independent.
  - The probability of success (p) is constant for each trial.

These can be modelled by binomial distribution.	These cannot be modelled by binomial distribution.
- The number of seeds germinating from a pack of 30 seeds planted in the same conditions. - The number of 6s in ten rolls of a dice. - The number of faulty lights in a box of 20	- The heights of 10 people (this is a *continuous* random variable, and there are not two possible outcomes). - The number of throws of a dice until a six occurs (there is not a fixed number of trials). - The number of red cards selected from a pack when four cards are drawn without replacement (probability of success is not constant since trials are not independent).

- $X \sim B(n, p)$ means X is binomially distributed with n trials and probability of success p
- The probability of r successes is: $P(X = x) = {}^nC_x p^x (1-p)^{n-x}$

For example, if X is binomially distributed with 6 trials and a probability of success of 0.4, you can write $X \sim B(6, 0.4)$

The probability of exactly four successes is given by:

$$P(X = 4) = {}^6C_4 \times 0.4^4 \times 0.6^2 = 0.1382$$

Calculator

Work out how to find cumulative binomial probabilities on **your** calculator:

Use DISTRIBUTION mode

Binomial CD
x	:2
N	:6
p	:0.4

Gives $P(X \le 2) = 0.5443$　　(give answers to 4 significant figures)

Given that $X \sim B(17, 0.54)$ you can use your calculator to find $P(X < 10)$

You will first need to use the fact that $P(X < 10) = P(X \le 9)$ since the binomial is a discrete distribution.

Then enter $x = 9$, $N = 17$ and $p = 0.54$ into your calculator to give

$$P(X < 10) = 0.5590$$

Example 1

The probability of a household purchasing semi-skimmed milk at least once in a week is estimated from a sample to be 72%.

a Explain why a binomial distribution is a suitable model for the number of households purchasing semi-skimmed milk in a sample of 12 households.

b Calculate the probability that semi-skimmed milk was bought by

    i Exactly 10 households,   **ii** More than 10 households.

The probability of a household purchasing cheese at least once in a week is estimated to be 34% and the probability of a household purchasing both cheese and semi-skimmed milk is estimated to be 32%.

c Are the events 'buying semi-skimmed milk' and 'buying cheese' independent?

Use probabilities to justify your answer.

d Calculate the probability that in a sample of 20 households, exactly half buy both cheese and semi-skimmed milk.

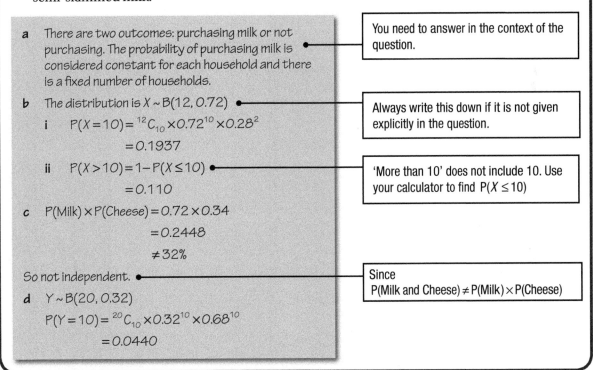

a There are two outcomes: purchasing milk or not purchasing. The probability of purchasing milk is considered constant for each household and there is a fixed number of households.

> You need to answer in the context of the question.

b The distribution is $X \sim B(12, 0.72)$

> Always write this down if it is not given explicitly in the question.

  i $P(X = 10) = {}^{12}C_{10} \times 0.72^{10} \times 0.28^2$

      $= 0.1937$

  ii $P(X > 10) = 1 - P(X \leq 10)$

      $= 0.110$

> 'More than 10' does not include 10. Use your calculator to find $P(X \leq 10)$

c $P(\text{Milk}) \times P(\text{Cheese}) = 0.72 \times 0.34$

                  $= 0.2448$

                  $\neq 32\%$

So not independent.

> Since
> $P(\text{Milk and Cheese}) \neq P(\text{Milk}) \times P(\text{Cheese})$

d $Y \sim B(20, 0.32)$

  $P(Y = 10) = {}^{20}C_{10} \times 0.32^{10} \times 0.68^{10}$

           $= 0.0440$

Exam tips

- Check that the conditions are met before using the binomial distribution.
- State the reasons why a binomial distribution is suitable in the context of the question.
- Always write down the distribution and parameters being used.
- Show full working even when using a calculator.
- Be careful with inequalities. For example, remember that $P(X < n) = P(X \leq n - 1)$ and that $P(a < X \leq b) = P(X \leq b) - P(X \leq a)$

1 Given that $X \sim B(8, 0.3)$, find to 3 dp

a $P(X = 2)$ [1]

b $P(X < 3)$ [1]

c $P(X \neq 5)$ [1]

2 Given that $X \sim B(6, 0.5)$, complete the following probability distribution table, giving your answers to 3 dp. [3]

x	0	1	2	3	4	5	6
$P(X = x)$							

3 The random variable T has a binomial distribution, $n = 10$, $p = \dfrac{1}{4}$
Find, giving your answers to 2 sf,

a $P(T = 3)$ [1]

b $P(T \geq 4)$ [1]

c $P(1 \leq T < 6)$ [1]

4 A fair six-sided dice is thrown seven times and the random variable X denotes the number of 4s obtained.

a Give the distribution of X [1]

b Find, giving your answers to 3 dp,

   i $P(X = 2)$ [1]

   ii $P(X > 2)$ [1]

   iii $P(0 \leq X < 3)$ [1]

5 A box contains five white, two red and three black balls. A ball is chosen at random.

This experiment is performed eight times with the ball replaced after each draw.

What is the probability that the ball is either red or white a total of six times? You should define a random variable and give its distribution. [3]

6 At a recycling plant, 43% of plastic items sent to the plant are suitable for recycling. Find the probability that, out of 25 plastic items considered, more than 10 are suitable for recycling. [2]

7 65% of English residents in 2013 had an iron intake of over 11 mg per day. Find the probability that, of 20 people sampled, more than 15 had an iron intake over 11 mg per day. State any assumptions you make in applying a particular distribution. [4]

8 A researcher is investigating the attitudes of people of different income groups towards purchasing fruit and vegetables. To complete his research, he must find five adults in the 9th decile income group.

a Write down the probability that and adult, chosen randomly from the entire population, will be in the required income group. [1]

b The researcher asks 30 of these randomly chosen adults about their income. What is the probability that he will find at least five in the 9th decile group? [2]

c Show that he should question 78 of the randomly chosen adults adults if he wants the probability of finding at least five in the 9th decile group to be at least 0.9 [3]

9 The average daily wind speed (in knots to the nearest integer) at a weather station for the 31 days in May 2015 were as follows.

5 6 6 7 7 7 7 8 8 8 8 8 9 9 9 9 9 9 10 10 10 10 11 12 12 13 13 14 14 18 19

a Use linear interpolation to show that the median wind speed for this period was $9\frac{1}{6}$ knots. [2]

b Write down the probability that, on a randomly chosen day, the mean wind speed will be above $9\frac{1}{6}$ knots. [1]

c Use the binomial distribution to find the probability that, in four randomly chosen days from May, there will be more than two days when the mean wind speed will be above $9\frac{1}{6}$ knots. [2]

d Comment on the validity of the assumption that the number of days on which the mean wind speed is above $9\frac{1}{6}$ knots follows a binomial distribution. [1]

10 It is estimated that 10% of children in Yorkshire buy over two litres of carbonated soft drink per week. A researcher needs to question at least 10 children who drink this amount per week.

a What is the probability that, if she asks 100 randomly selected children from Yorkshire about the quantity of soft drink they buy, she will obtain the required sample? [3]

b How many children should she question so that she has a probability of at least 0.95 of obtaining the required sample? [3]

Recap

- A **hypothesis test** assumes that the **null hypothesis**, H_0, is true and examines if there is sufficient evidence to reject it in favour of the **alternative hypothesis**, H_1
- You can test the probability of success in a binomial distribution: the null hypothesis will be
  - $H_0: p = p_1$
- The alternative hypothesis can be either
  - $H_1: p > p_1$ or $H_1: p < p_1$ for a **one-tailed test**
  - $H_1: p \neq p_1$ for a **two-tailed test**.

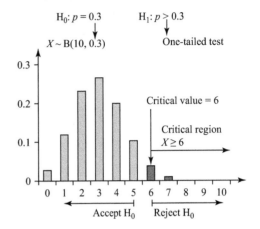

$H_0: p = 0.3$ $H_1: p > 0.3$

$X \sim B(10, 0.3)$ One-tailed test

Critical value = 6

Critical region $X \geq 6$

Accept H_0 Reject H_0

To test whether a coin is fair, you can use a two-tailed or a one-tailed test.
Let p = the probability of 'heads'

A two-tailed test would have hypotheses

$H_0: p = \dfrac{1}{2}$ and $H_1: p \neq \dfrac{1}{2}$

If you suspected that the coin was biased towards 'tails', a one-tailed test would have hypotheses

$H_0: p = \dfrac{1}{2}$ and $H_1: p < \dfrac{1}{2}$

- A **test statistic** is a value of the random variable found from a sample.
- The **critical region** is the set of values of the random variable that are sufficiently unlikely to occur under the null hypothesis.
  Precisely how unlikely is determined by the given **significance level**.
- The **critical value** is the boundary of the critical region.
  The critical value lies just inside the critical region.
- If the test statistic is inside the critical region, then you reject the null hypothesis.
- The region outside the critical region is called the **acceptance region**.
- If the test statistic is inside the acceptance region, then you accept the null hypothesis.

Suppose you wish to perform a one-tailed test at the 5% significance level to see if a coin is biased towards tails.

- Flip the coin 30 times and count the number of heads, H – this is your test statistic.
- Assuming that the coin is fair (the null hypothesis),
  the probability of getting 10 or fewer 'heads' is 0.0494 = 4.94%
  but the probability of getting heads 11 or fewer times is 0.1002 > 5%
- Therefore the critical region for the test statistic is $H \leq 10$
- That is, the critical value is 10
- If your test statistic is in the region $H \leq 10$, then you can reject the null hypothesis at the 5% significance level as there would be evidence to suggest that the coin is biased towards 'tails'.

Assuming the null hypothesis, H_0, is true, then the significance level tells you how likely you are to *incorrectly* reject H_0 when applying the test. As you decrease the significance level, the critical region gets smaller, and this error is less likely to occur. That is, if you do reject H_0, it is more likely to be because H_0 is not true than you have been unlucky.

 **MyMaths**   🔍 2115   SEARCH

Example 1

The probability that a randomly chosen household in the UK has bought butter in any given week is assumed to be 70%. A shopkeeper in a town in the South West claims that the probability is higher in his area. He samples 50 households and 40 of them have bought butter in the last week. The critical value at the 5% significance level is 41

a For this test state

   i The hypotheses,   **ii** The test statistic,

   iii The critical region,   **iv** The acceptance region.

b What conclusions can you draw from the test statistic?

> **a** **i** $H_0: p = 0.7$ $H_1: p > 0.7$
>
>   **ii** The test statistic is 40
>
>   **iii** $X \geq 41$ is the critical region
>
>   **iv** $X \leq 40$ is the acceptance region
>
> **b** 40 lies in the acceptance region, and therefore we do not reject the null hypothesis.
> There is insufficient evidence at the 5% significance level to support the shopkeeper's claim.

This is a one-tailed test; you are testing if the probability is higher than 70%

Remember that you need 41, *or a more extreme result,* for the critical region.

Always make a comment using the context of the question.

Example 2

A hypothesis test is set up with $H_0: p = \dfrac{1}{4}$ and $H_1: p \neq \dfrac{1}{4}$

The critical values at the 10% significance level are $X = 2$ and $X = 9$

a State the critical region for this test.

b What effect will lowering the significance level have on the probability of incorrectly rejecting the null hypothesis?

> **a** $X \leq 2$ and $X \geq 9$
>
> **b** The probability of incorrectly rejecting the null hypothesis will *decrease* in line with the significance level.

This is a two-tailed test, so there are two parts to the critical region. Remember that you need to include the critical values in the critical region.

Exam tips

One-tailed and two-tailed tests

- In a two-tailed test, the significance value is split in half so that half is at the top of the distribution and half at the bottom. Thus, to do a 10% two-tailed test, you need 5% at the top and bottom.
- Setting up the alternative hypothesis is important in deciding what sort of test to use.
  - If it is thought that the claim is simply wrong, then the alternative hypothesis will be $x \neq$ 'a value' and a two-tailed test should be used.
  - If it is thought to be too high or too low, then the alternative hypothesis will involve an inequality and a one-tailed test should be used.

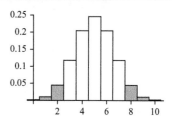

The critical region for a two-tailed test.

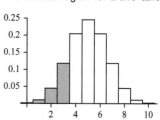

The critical region for a one-tailed test.

1   A random variable X can be modeled by a binomial distribution such that $X \sim B(40, p)$
    A one-tailed test has hypotheses $H_0: p = p_1$ and $H_1: p > p_1$
    At the 5% significance level, the critical region for the test is $X > 24$
    The same test is repeated with the significance level changed to 10%

    What is the new critical region?   [1]

    Circle your answer.

    $X > 23$ $X > 24$ and $X < 16$ $X > 12$ $X < 12$

2   A study into peoples' attitude to crime in 2006 showed that 64% of people thought that national crime
    rates were increasing. When questioned about *local* crime, an independent random sample of 45 adults
    included 20 people who agreed that it was increasing.

    a   State the null and alternative hypotheses for a test to find out whether the view on local crime was
        different from that on national crime.   [2]

    _____

    _____

    _____

    _____

    b   Let X be a random variable for the number of people in the sample of 45 adults who agree with the
        proposition that local crime is on the increase.

        Give the distribution of X under the null hypothesis defined in part **a**.   [2]

    _____

    _____

    _____

    _____

3 It has been estimated that, in response to increased food prices between 2007 and 2013, 22% of households reacted by 'trading down', that is, buying an alternative cheaper product. A sample of 40 adults were asked about their response to food price increases and 14 accepted trading down as the main response.

   a State a condition on the method of choosing the sample so that a binomial probability model can be used to test the estimate. [1]

   b Assuming that the condition in part **a** is met, write down the null and alternative hypotheses in a test of this estimate. [2]

   Let the random variable X = the number of households out of 40 who trade down.

   c Give the distribution of X, assuming the null hypothesis to be true. [2]

   d Under the null hypothesis, at a significance level of 5%, the critical value is 14. State the conclusion of the test. [1]

4 a Explain the term 'critical region' in the context of hypothesis testing and state why an observation in the critical region suggests that the null hypothesis should be rejected. [2]

4 b Between 2010 and 2015, it is believed that the sales of white bread in England fell significantly. In 2010, the median amount of white bread bought was 280 g per person per week. To test the belief that sales were falling, in 2015 researchers took a sample of 16 adults from England and found that 4 bought more than 280 g of white bread per week.

Let X be the number in the sample who bought more than 280 g of white bread per week.

i State the null and alternative hypotheses for the required test and give the distribution of X, assuming that the null hypothesis is true. [2]

_____

_____

_____

ii Under the null hypothesis, at a significance level of 5%, the critical value is 4. Write down the result of the test. [1]

_____

_____

5 a State the conditions under which a random experiment can be modelled by a binomial probability distribution.
You should clearly define the relevant random variable. [4]

_____

_____

_____

_____

_____

5 **b** State the probability distribution function, $P(X=x)$, for a random variable X which has a binomial distribution with parameters n and p [3]

_____

_____

_____

_____

6 In 2010, the proportion of individuals in England purchasing more than 380 ml of full fat liquid milk was found to be 24%. In a sample of 28 individuals taken in 2014, 3 purchased more than this amount. A statistical test is to be carried out to use this statistic to determine whether the amount spent on this type of milk has decreased over this period.

a State a condition for the sample to be suitable for use in the test and state why the condition is necessary. [2]

_____

_____

_____

_____

Let X be a random variable for the number of individuals in the sample who purchase more than 380 ml of full fat liquid milk.

b State the hypotheses H_0 and H_1 and give the distribution of X [3]

_____

_____

_____

_____

c At a significance level of 10%, the critical value is 3. State the conclusion of the hypothesis test. [1]

_____

_____

Recap

- To find the critical region, you assume the null hypothesis in order to calculate probabilities.
- The critical value lies inside the critical region.
- For the binomial distribution, $X \sim B(n, p)$, it is very unlikely that there will be an integer, m, such that $P(X \geq m)$ exactly equals the required significance level.

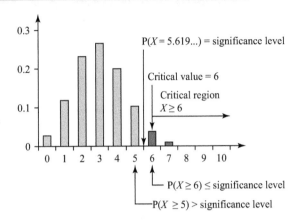

  In practice, for a high-values, in a one-tailed test, you find the *smallest m* such that
  $$P(X \geq m) \leq \text{significance level}$$
  $$[\text{or } P(X \geq m - 1) > \text{significance level}]$$

  The critical region is $X \geq m$
- For low values you find the *largest* value of m' such that
  $$P(X \leq m') \leq \text{significance level}$$
  $$[P(X \geq m' + 1) > \text{significance level}]$$

If you have $H_0: p = 0.1$ and $H_1: p < 0.1$ and you wish to find the critical region at the 5% significance level, when there are 60 trials, you must find the largest value of x such that
$P(X \leq x) \leq 0.05$

You need to assume that $X \sim B(60, 0.1)$

By adding up the probabilities, you can see that

$P(X \leq 1) \leq 0.0138$ which is less than 5%

$P(X \leq 2) \leq 0.0530$ which is greater than 5%

The critical value is 1 and the critical region is $X \leq 1$

Create a table on your calculator using the function

$f(x) = {}^{60}C_x \times 0.1 \times 0.9^{60 - x}$

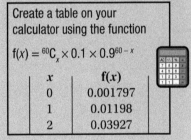

x	$f(x)$
0	0.001797
1	0.01198
2	0.03927

- Alternatively, you can carry out a hypothesis test by finding the **p-value** of a test statistic; this is the probability of obtaining the test statistic or *worse* under the null hypothesis.

Suppose $H_0: p = 0.6$ and $H_1: p \neq 0.6$ and that out of 10 trials, 9 were successful.

With 10 trials, the expected number of successes is $10 \times 0.6 = 6$, so 9 is more than expected.
Hence you need to calculate the probability of getting 9 or *more* successes out of 10 trials.

$P(X \geq 9) = 0.0403 + 0.0065$
$\qquad\quad = 0.046$

In this case, the p-value is 0.0468 (4.68%)

$H_1: p \neq 0.6 \Rightarrow$ use a two-tailed test: you split the 5% between the two tails.
Therefore you need to compare the p-value to 2.5%

Create a table on your calculator using the function

$f(x) = {}^{10}C_x \times 0.6^x \times 0.4^{10 - x}$

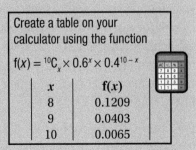

x	$f(x)$
8	0.1209
9	0.0403
10	0.0065

$4.68\% > 2.5\% \Rightarrow$ do not reject the null hypothesis.
There is insufficient evidence to assume that the probability of success is not 0.6

Example 1

A supermarket sells eggs in boxes of 6. The probability of an egg being broken is 1%. A crate of 20 boxes of eggs contains 3 broken eggs. Assuming that the number of broken eggs can be modelled by a binomial distribution, use a hypothesis test with a 10% significance level to test whether this an unusually large number of eggs to be broken.

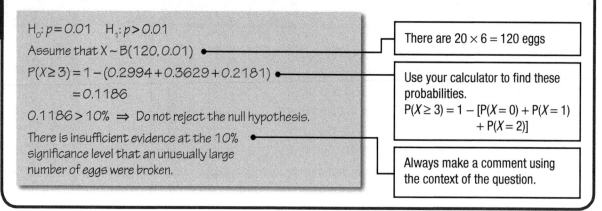

$H_0: p = 0.01$ $H_1: p > 0.01$

Assume that $X \sim B(120, 0.01)$

$P(X \geq 3) = 1 - (0.2994 + 0.3629 + 0.2181)$

$\qquad = 0.1186$

$0.1186 > 10\% \Rightarrow$ Do not reject the null hypothesis.

There is insufficient evidence at the 10% significance level that an unusually large number of eggs were broken.

There are $20 \times 6 = 120$ eggs

Use your calculator to find these probabilities.
$P(X \geq 3) = 1 - [P(X = 0) + P(X = 1) + P(X = 2)]$

Always make a comment using the context of the question.

Example 2

The probability of a particular bus being on time has been found to be 52%
The bus company wants to test whether this has changed.

a   Write down the null and the alternative hypotheses.

A binomial model is to be used and a random sample of 11 buses will be taken.

b   Find the critical region at the 5% significance level.

c   What is the probability of incorrectly rejecting the null hypothesis in this case?

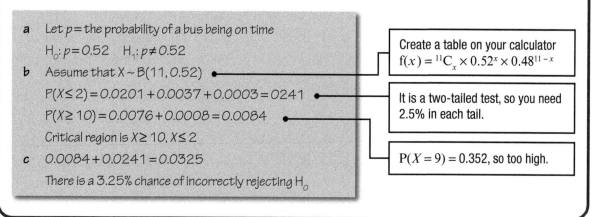

a   Let $p =$ the probability of a bus being on time

$H_0: p = 0.52$ $H_1: p \neq 0.52$

b   Assume that $X \sim B(11, 0.52)$

$P(X \leq 2) = 0.0201 + 0.0037 + 0.0003 = 0.0241$

$P(X \geq 10) = 0.0076 + 0.0008 = 0.0084$

Critical region is $X \geq 10$, $X \leq 2$

c $0.0084 + 0.0241 = 0.0325$

There is a 3.25% chance of incorrectly rejecting H_0

Create a table on your calculator
$f(x) = {}^{11}C_x \times 0.52^x \times 0.48^{11-x}$

It is a two-tailed test, so you need 2.5% in each tail.

$P(X = 9) = 0.352$, so too high.

Exam tips

Calculating critical values

In the exam AQA will not provide tables of cumulative binomial distributions so you must learn how to find critical values using *your own* calculator. If your calculator can find inverse binomial distributions, you should check if it gives the value of x such that $P(X \leq x)$ is closest to the required significance level. This may mean that $P(X \leq x)$ is slightly larger than the required significance level and that the critical value is $x - 1$. The correct value can be found by inspecting the values of $P(X \leq x)$ and $P(X \leq x - 1)$

For example, suppose that $X \sim B(20, 0.3)$ and a 10% significance level is required for a low-values, one-tailed test. A calculator may give $x = 3$ as $P(X \leq 3) = 10.7...\%$, whereas the correct critical value is $x = 2$ for which $P(X \leq 2) = 3.54...\%$

n	$P(X \leq n)$
0	0.000798
1	0.007637
2	0.035483
3	0.107087
4	0.237508
5	0.416371

1 **a** A random variable X has a binomial distribution, $X \sim B(6, p)$. The value of p was known to be 0.72 but is now believed to have decreased.

What are the null and alternative hypotheses in a test of this belief? [1]

Circle your answer.

      A $H_0: p = 0.72$ $H_1: p \neq 0.72$ **B** $H_0: p = 0.72$ $H_1: p < 0.72$

      C $H_0: p < 0.72$ $H_1: p = 0.72$ **D** $H_0: p = 0.72$ $H_1: p < 0.19$

b The table below gives the probabilities, to 2 significant figures, of all values of X assuming that the null hypothesis is true.

x	0	1	2	3	4	5	6
$P(X = x)$	0.00048	0.0074	0.048	0.16	0.32	0.33	0.14

Using a significance level of 5%, what values of X would suggest that the belief is correct?

Explain your answer. [2]

2 In 2013, 70% of adults had a higher than recommended percentage of calorie intake derived from saturated fatty acids. Two years later, it was claimed that this proportion had increased. To test this, a random selection of 16 adults were investigated and 14 were found to have a higher than recommended percentage of calorie intake derived from saturated fatty acids.

a Perform a hypothesis test of this claim using a significance level of 10%

You should state clearly the null and alternative hypotheses. [5]

2 **b** Comment on the reliability of the test. [1]

_____

_____

_____

_____

3 In 2016, a large investigation found that 6 out of every 10 people questioned agreed that the quantity of butter they bought each week had increased over the last year. This year, in an independent, random sample of 25 people, 20 people agreed with this proposition. A test was carried out to determine whether the proportion of people agreeing with the proposition had changed within the whole population.

a Write down the null and alternative hypotheses for the test. [2]

_____

_____

_____

_____

b Perform the test at a significance level of 5%, stating clearly your conclusion. [3]

_____

_____

_____

_____

_____

_____

4   In 2010, the median amount of soft drinks bought was 1206 ml per person per week. In order to test whether this had changed by 2014, 35 people were asked about the amount of soft drink they had bought in one week. Twelve of them had bought less than 1206 ml in that week.

Test, at a significance level of 5%, whether this is evidence of a change in the amount of soft drink bought during the four years. [5]

5   A chicken farm has a hatch rate (the percentage of fertilized eggs that hatch after incubation) of 52%. Following the introduction of new incubation equipment, a random sample of 64 eggs produced 39 chicks.

a   Explain why the binomial probability distribution could provide a good model for the number of chicks produced from these eggs. [3]

5 **b** Complete the following table for X, a random variable for the number of chicks from a total of 64 eggs using the *old* equipment. [3]

x	39	40	41	≥ 42
P($X = x$)				

c It is claimed that the new equipment has increased the hatch rate for fertilized eggs.

Test, using a significance level of 5%, whether the claim is justified. [2]

_____

_____

_____

_____

6 In the first half of the 18th century, the mean height of an adult English male was 165 cm. A comparative study on the heights of adult Irish males found 27 with heights over 165 cm in a sample of 40 randomly chosen Irish males.

a Explain why it is reasonable to assume that the probability that a randomly chosen English male exceeds 165 cm is 0.5 [2]

_____

_____

_____

b Let X be the number in the sample of 40 Irish males with heights over 165 cm.

Using a significance level of 2%, test the belief that Irish males are on average taller than English males. [2]

_____

_____

_____

_____

Hypothesis testing The critical region

Recap

- In set notation:
  - $A \cup B$ is the **union** and means A or B
  - $A \cap B$ is the **intersection** and means A and B
  - A' is the **complement of A** and means *not A*
  - $A|B$ is a **conditional** statement and means *A given B*

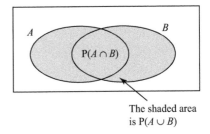

The shaded area is $P(A \cup B)$

- The addition rule for probability states that
  $$P(A \cup B) = P(A) + P(B) - P(A \cap B)$$

If the probability of event A occurring is 0.4, the probability of event B is 0.7 and the probability of both occurring is 0.15. You can draw a Venn diagram to illustrate these probabilities.

Using the Venn diagram you can see that

$$P(A \cup B) = 0.25 + 0.15 + 0.55 = 0.95$$

Check using the addition rule:

$$P(A \cup B) = 0.7 + 0.4 - 0.15 = 0.95 \text{ as expected.}$$

This is 0.7 – 0.15

Remember that the probabilities must add up to 1

Complete this first

- The probability of event A occurring *given* event B has occurred is given by
  $$P(A|B) = \frac{P(A \cap B)}{P(B)}$$

- If events A and B are independent then they have no effect on each other. The fact that B has occurred does not affect the probability of A happening and so
  $$P(A|B) = P(A)$$

If events A, B and C are such that $P(A) = \frac{1}{6}$, $P(B) = \frac{2}{5}$, $P(C) = \frac{3}{4}$ and $P(A \cap B) = \frac{1}{10}$, $P(A \cap C) = \frac{1}{8}$

Then the conditional probability of A given B is

$$P(A|B) = \frac{P(A \cap B)}{P(B)}$$

$$= \frac{\frac{1}{10}}{\frac{2}{5}}$$

$$= \frac{1}{4}$$

The conditional probability of C given A is

$$P(C|A) = \frac{P(A \cap C)}{P(A)}$$

$$= \frac{\frac{1}{8}}{\frac{1}{6}}$$

$$= \frac{3}{4}$$

This is the same as $P(C)$; therefore events A and C are independent.

Example 1

The event M is a household purchasing infant milk and the event F is a household purchasing baby food. Using data from a sample, the probabilities are estimated to be

$P(M) = 0.01$, $P(F|M) = 0.25$ and $P(F|M') = 0.05$

a Draw a tree diagram to show this information.

b Calculate the probability of a household purchasing baby food.

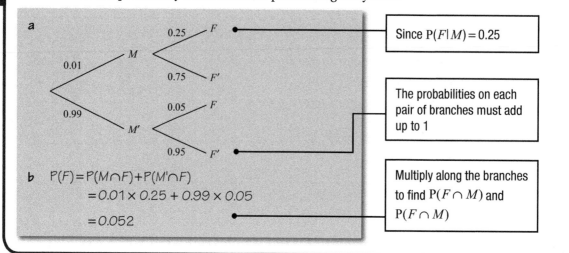

a

b $P(F) = P(M \cap F) + P(M' \cap F)$

$\quad\quad = 0.01 \times 0.25 + 0.99 \times 0.05$

$\quad\quad = 0.052$

Since $P(F|M) = 0.25$

The probabilities on each pair of branches must add up to 1

Multiply along the branches to find $P(F \cap M)$ and $P(F \cap M)$

Example 2

Calculate the probability of B not occurring when

$P(A) = 0.3$, $P(A|B) = 0.5$, $P(A \cup B) = 0.7$

$P(A \cap B) = 0.3 + P(B) - 0.7$

$0.5 = \dfrac{0.3 + P(B) - 0.7}{P(B)}$

$0.5P(B) = 0.3 + P(B) - 0.7 \Rightarrow P(B) = 0.8$

Therefore, $P(B') = 1 - 0.8 = 0.2$

Using addition rule:
$P(A \cap B) = P(A) + P(B) - P(A \cup B)$

Using conditional probability rule:
$P(A|B) = \dfrac{P(A \cap B)}{P(B)}$

Exam tips

- It is usually a good idea to draw a diagram to illustrate the situation. A Venn diagram or a tree diagram can often help you see what is going on.
- If you are asked to justify whether events are independent or mutually exclusive, then you must use probabilities.
- To show that A and B are independent, verify any one of:
  - $P(A \cap B) = P(A) \times P(B)$
  - $P(A) = P(A|B)$
  - $P(B) = P(B|A)$
- To show that A and B are mutually exclusive, verify any one of:
  - $P(A \cap B) = 0$
  - $P(A \cup B) = P(A) + P(B)$
- Remember that probabilities must be between 0 and 1, so go back and check your working if you get an answer outside of this range.

1   Two boxes each contain 3 white balls and 5 red balls. A ball is chosen at random from box number one and placed in box number two. A ball is then drawn from box two.

a   Draw a tree diagram showing all possible outcomes for this experiment.   [2]

b   Use your tree diagram to find the probability that the ball selected from box two is red.   [2]

2   50 adults were asked to calculate their body mass index (BMI) using $BMI = \dfrac{\text{weight in kg}}{(\text{height in m})^2}$ and also to record how much sugar they bought over a period of one week. The results are given in the table.

BMI / Sugar bought	< 25	25–30	> 30	Total
≥ population median	3	10	15	28
< population median	8	6	8	22
Total	11	16	23	50

a   Find the probability that a person chosen at random

i   Buys less than the median amount of sugar,   [1]

ii   Has a BMI of at least 25 given that the amount they spend on sugar is less than the median.   [2]

b By considering people whose BMI is less than 25, people who buy less than the median amount of sugar per week or both, decide whether it is likely that BMI and sugar consumption are independent of each other. [3]

3 Integers 1 to 21 are classified as   P: prime,   E: even,   N: greater than 10

   a Draw a Venn diagram showing the number of integers in each section of the diagram. [1]

   b One of the numbers is chosen at random. Find the probability that the number

     i Is prime, even and greater than 10, [1]

     ii Is not prime, not even and not greater than 10, [1]

     iii Is even, given that it is greater than 10 [1]

4 A large packet of mixed poppy seeds contains three varieties, A, B and C, in the ratio $1:4:5$. The germination rates of the three varieties are 67%, 79% and 89%, respectively.

A seed is chosen at random from the packet and G is the event: the seed germinates.

   a Write down $P(G|A)$ [1]

   b Find $P(G \cap A)$ [2]

c Find P(*G*) [3]

_____

_____

_____

d Find P(*A*|*G*) [2]

_____

_____

5   In an investigation into regional food purchasing patterns, a large group of English adults were questioned about the amount of confectionary they bought per week. The proportions of adults from three regions of England and the probability that the quantity they bought exceeded the national median were recorded in this table.

Region	London	South West	North East
Proportion of sample	0.32	0.39	0.29
Probability of exceeding national median	0.31	0.54	0.59

A person is chosen at random. Find the probability that

a The person is from London and exceeds the national median amount, [2]

_____

_____

_____

b The person is from London given that he/she exceeds the national median amount. [5]

_____

_____

_____

_____

_____

_____

Recap

Real-life situations can be modelled using discrete distributions.

- The discrete uniform distribution can be used when:
  - There is a fixed number of outcomes (n)
  - The probability of each outcome is the same $\left(p = \dfrac{1}{n} \right)$

- The binomial distribution can be used to model a real-life situation if:
  - There are only two possible outcomes (e.g. success/failure)
  - There are a fixed numbers of trials (n)
  - The trials are independent
  - The probability of success (p) is constant for each trial.

Discrete uniform model	Binomial model
A game involves players choosing a number from 1 to 8. If each number is equally likely to be chosen, then the outcome can be modelled by a discrete uniform distribution with probability $\dfrac{1}{8}$ for each number.	Joe went swimming 73 times last year. The probability that he went swimming on any chosen day is $\dfrac{73}{365} = \dfrac{1}{5}$ So the number of times he goes swimming in a week could be modelled by $X \sim B\left(7, \dfrac{1}{5} \right)$
In practice, it may be that each number is not equally likely to be chosen, in which case, a table could be drawn to show the actual probability distribution.	A possible problem with this model is that the trials may not be independent. For example, if he goes swimming on Monday, this might mean he is less likely to go swimming on Tuesday.

You can use data to estimate the parameters of a binomial model.

Flora drops her slice of toast five times and it lands butter-side down on three occasions. She thinks that the number of times it lands butter-side down can be modelled by a binomial distribution since there are two possible outcomes, 'butter-side down' or 'butter-side up', and she thinks it is reasonable to assume that the probability of it landing 'butter-side down' is constant and independent with each trial.

In order to find a more accurate estimation of the probability, she asks some friends to repeat the experiment and records all the results in a table.

Number of times toast lands butter-side down, X	0	1	2	3	4	5
Frequency	1	2	3	10	7	5

The mean is $\bar{x} = \dfrac{(0 \times 1) + (1 \times 2) + (2 \times 3) + (3 \times 10) + (4 \times 7) + (5 \times 5)}{28} = 3.25$.

On average, the toast lands butter-side down 3.25 times in every 5 trials so a suitable value of p could be $\dfrac{3.25}{5} = 0.65$

Therefore the number of times her toast lands butter-side down, X, can be modelled as $X \sim B(5, 0.65)$

Example 1

The numbers of days per week that Simon eats chocolate is modelled by a binomial distribution with $p = 0.35$

a For a randomly chosen week, calculate the probability that

i Simon does not eat any chocolate,

ii Simon eats chocolate on more than four days.

b Explain what assumptions must be made in order to use this model.

Simon records the number of days per week that he eats chocolate over a 24-week period and collates the results in the table.

Number of days	0	1	2	3	4	5	6	7
Frequency	3	3	2	5	3	6	2	0

c Use the data to suggest a better value for p in the model.

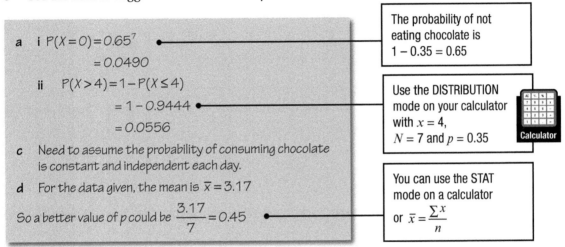

a **i** $P(X=0) = 0.65^7$

        $= 0.0490$

The probability of not eating chocolate is $1 - 0.35 = 0.65$

ii $P(X>4) = 1 - P(X \leq 4)$

        $= 1 - 0.9444$

        $= 0.0556$

Use the DISTRIBUTION mode on your calculator with $x = 4$, $N = 7$ and $p = 0.35$

c Need to assume the probability of consuming chocolate is constant and independent each day.

d For the data given, the mean is $\bar{x} = 3.17$

So a better value of p could be $\dfrac{3.17}{7} = 0.45$

You can use the STAT mode on a calculator or $\bar{x} = \dfrac{\sum x}{n}$

Exam tips

- When asked what assumptions must be made, always give your answer in the context of the question.
- Show full working even when using a calculator.
- Be careful with inequality signs, especially when you're using your calculator to find binomial probabilities.
- Remember that you may need to use your knowledge of the Large data set so ensure that you are familiar with it.

1   A fashion outlet monitors its monthly sales figures for the first half of 2015 and wishes to find out whether sales are spread uniformly throughout the period. To test this, the total till receipts for January to June are noted. The results are shown in the table.

Month	Jan	Feb	Mar	Apr	May	June
Receipts (£000s)	15.3	13.6	12.3	14.1	15.8	17.2

a   Write down the expected sales figures, assuming that sales are spread uniformly over this period. Your distribution should have the same total as the data above.   [2]

Month	Jan	Feb	Mar	Apr	May	June
Receipts (£000s)						

b   By comparing the two distributions, comment on the belief that sales are uniformly distributed.   [2]

2   The median amount of bread bought per person per week for the ten years from 2005 is shown for a sample of English adults:

Year	2005	2006	2007	2008	2009	2010	2011	2012	2013	2014
Amount of bread (g)	688	679	663	646	641	618	609	603	595	549

a   Complete the following table to show expected bread sales for 2005–2014, assuming that the total was unchanged and spread uniformly throughout the 10 years.   [2]

Year	2005	2006	2007	2008	2009	2010	2011	2012	2013	2014
Amount of bread (g)										

b   Does the data suggest that bread purchasing has remained constant over these ten years? Give two reasons to support your answer.   [2]

3 A bag contains red, blue and black counters. It is believed that these occur in the ratio $1:2:5$

   a A counter is drawn from the bag. Complete the following table of the probabilities for each outcome, assuming this belief to be correct. [2]

Colour	Red	Blue	Black
Probability			

   This experiment is repeated a total of 60 times and gives the following results.

Colour	Red	Blue	Black
Frequency	7	18	35

   b Do these results support the suggested ratio? Give reasons for your answer. [2]

_____

_____

_____

_____

4 **a** An experiment consists of a fixed number of trials, each of which can result in one of two possible outcomes. Give two reasons why, when the experiment is performed, the number of times one of the outcomes occurs might NOT follow a binomial distribution. [2]

_____

_____

_____

_____

   b A random variable X is believed to follow a binomial distribution with $n = 6$, $p = 0.4$

   Complete the following table, assuming that this belief is true. Give your answers to 3 dp. [2]

_____

_____

X	0	1	2	3	4	5	6
$P(X = x)$							

   c A sample of 100 values of X is taken and gives the following results.

X	0	1	2	3	4	5	6
Frequency	0	1	6	19	32	30	12

Does the data support this belief? If not, without further calculation, comment on how the model should be changed to improve the fit of observed to expected frequencies. [3]

5   In 2014, a newspaper headline stated: 'Mineral water sales almost double to 350 ml per week over 15 years'. In 2016, to investigate whether sales have continued to increase, 50 people were asked how much mineral water they had bought over the last week. 33 of these had bought more than 350 ml.

   a   Write down the distribution of X, the number of people who bought more than 350 ml of mineral water, assuming that there had been no increase on the 2014 figure. State any assumptions you make. [2]

   b   Calculate the mean of X [1]

   c   Using the distribution in part **a**, find the probability that X takes a value of at least 33 [1]

   d   Use your answers to parts **b** and **c** to decide whether sales of mineral water are likely to have increased. Give reasons for your answer. [2]

Recap

Continuous random variables can often be modelled by a **Normal distribution**.

- If X is a Normally distributed random variable with mean μ and standard deviation σ, then you write $X \sim N(\mu, \sigma^2)$
- The probability distribution function of X has a bell-shaped curve.
- The line of symmetry of the curve is the mean, μ
- The points of inflection on the curve are at the points $x = \mu \pm \sigma$
- The **standard Normal distribution**, Z, has mean $\mu = 0$ and standard deviation $\sigma = 1$, so you write $Z \sim N(0, 1^2)$
- You should use your calculator to find Normal probabilities.

P(x)

$\mu - \sigma \quad \mu \quad \mu + \sigma$ x

If you have a Normally distributed random variable, X, with mean 5 and standard deviation 3, then write $X \sim N(5, 9)$

Be careful: remember that it is the *variance* which is given as a parameter and $3^2 = 9$

Use you calculator to see that

$P(3 < X < 8) = 0.2185$

On your calculator:

Use DISTRIBUTION mode, and find Normal CD.

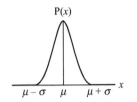
Calculator

Lower : 3
Upper : 8
σ : 3
μ : 5

P = 0.5889

Now use you calculator to see that

$$P(X > 4) = 0.5442$$

Since X is a *continuous* random variable, there is no difference between $P(X > 4)$ and $P(X \geq 4)$

On your calculator:

The upper limit is ∞, so if your calculator requires you to enter a value here just choose any really big number, e.g.

Calculator

Lower : 4
Upper : 999
σ : 3
μ : 5

P = 0.6306

Still using $X \sim N(5, 9)$, you can use your calculator to find the value of a such that $P(X > a) = 0.3$
The calculator gives you x such that
$P(X < x) =$ area, so you need to find a such that
$P(X < a) = 1 - 0.3 = 0.7$, which is
$\qquad a = 9.7196$

On your calculator:

Use DISTRIBUTION mode, and select Inverse Normal.

Calculator

Area : 0.3
σ : 3
μ : 5

xInv = 6.5732

- If you do not know the values of μ or σ, you will need to use the rule $Z = \dfrac{X - \mu}{\sigma}$ and then calculate with the standard Normal distribution on your calculator. There is an example of this on the next page.

Example 1

The amount of brown bread purchased per person in a sample from the South West is modelled by a Normal distribution with mean 170 g and variance 90

Calculate the probability that a randomly chosen household purchases less than 160 g of brown bread per person.

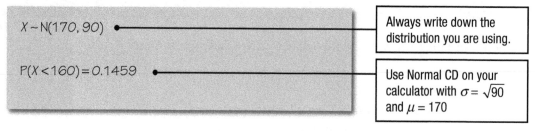

$X \sim N(170, 90)$ •————————— Always write down the distribution you are using.

$P(X < 160) = 0.1459$ •————————— Use Normal CD on your calculator with $\sigma = \sqrt{90}$ and $\mu = 170$

Example 2

A continuous random variable, X, is Normally distributed with mean μ and standard deviation σ. Calculate the values of μ and σ to 3 significant figures, given that $P(X > 8) = 20\%$ and $P(X < 1) = 5\%$

This question requires you to solve simultaneous equations.

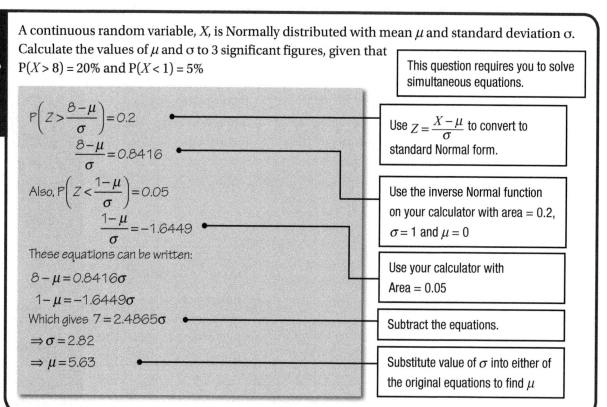

$$P\left(Z > \frac{8 - \mu}{\sigma}\right) = 0.2$$

$$\frac{8 - \mu}{\sigma} = 0.8416$$

Also, $P\left(Z < \frac{1 - \mu}{\sigma}\right) = 0.05$

$$\frac{1 - \mu}{\sigma} = -1.6449$$

These equations can be written:

$$8 - \mu = 0.8416\sigma$$

$$1 - \mu = -1.6449\sigma$$

Which gives $7 = 2.4865\sigma$

$$\Rightarrow \sigma = 2.82$$

$$\Rightarrow \mu = 5.63$$

Use $Z = \frac{X - \mu}{\sigma}$ to convert to standard Normal form.

Use the inverse Normal function on your calculator with area = 0.2, $\sigma = 1$ and $\mu = 0$

Use your calculator with Area = 0.05

Subtract the equations.

Substitute value of σ into either of the original equations to find μ

Exam tips

- Make sure your answer is written using clear notation.
- Be careful not to mix up the standard deviation and the variance.
- Use full accuracy for intermediate values and only round your final answer.
- You might find it helpful to sketch a graph of the distribution, showing the probability.

1 If $Z \sim N(0, 1)$, find the probability that

   a $P(Z < 1.2)$ [1]

   b $P(1.2 < Z < 1.9)$ [1]

   c $P(-0.3 < Z < 1.4)$ [1]

2 The variable X has the distribution $X \sim N(14, 5^2)$. By converting to a standard Normal variable Z, find the probability that

   a $P(X < 18)$ [1]

   b $P(10 < X < 20)$ [1]

   c $P(12 < X < 19)$ [1]

3 The random variable, X, has mean 7 and standard deviation 2. Find the value of a given that
$P(a < X < 9) = 0.61$ [3]

4 In 2014, the mean amount of sugar and preserves bought in England was 109 g per person per week.

   a Assuming this quantity to be Normally distributed with standard deviation of 12.4 g, find the probability that a randomly chosen person bought between 100 g and 120 g.   [2]

Five people are chosen at random.

b Find the probability that at least two of them bought between 100 g and 120 g. [2]

_____

_____

_____

5 a Show that, for a Normally distributed population with mean μ and variance σ^2, approximately 95% of observations lie within the interval $\mu \pm 2\sigma$ [3]

_____

_____

_____

b The mass of cheese, m grams, bought by 60 individuals in London during one week, is shown in the table.

i Calculate the mean and standard deviation of this distribution. [2]

Mass, m (g)	Frequency
$0 \leq m < 30$	6
$30 \leq m < 60$	12
$60 \leq m < 80$	15
$80 \leq m < 100$	11
$100 \leq m < 120$	8
$120 \leq m < 150$	8

_____

_____

ii Using linear interpolation, find the proportion of observations that lie within two standard deviations of the mean. [3]

_____

_____

_____

_____

c Use your answers to parts **a** and **b ii,** and any other information, to decide whether the quantity of cheese bought per week is likely to be Normally distributed. [2]

_____

_____

_____

_____

6 200 individuals in the East Midlands were asked to record the amount of butter they bought in one week. The amount, X, in grams, gave the following statistics.

Smallest value: 18 g   Largest value: 64 g

$\Sigma x = 7810$ $\Sigma x^2 = 336\,101$

a i Find the mean, $\bar{x}$, and standard deviation, σ, of X [2]

ii Hence find the values of $\bar{x} \pm 3\sigma$ [1]

b Use your answer to part **a ii**, and any other given information, to explain why it is reasonable to assume that X follows a Normal distribution. [2]

c Use the values in part **a i** to calculate the probability that a randomly chosen person from the East Midlands buys more than 50 g of butter in one week. [1]

7 Find the unknown parameters given the following information.

a $X \sim N(\mu_X, 9)$; $P(X < 10) = 0.7$ [3]

b $Y \sim N(\mu_Y, \sigma_Y^2)$; $P(Y < 12) = 0.7, P(Y < 15) = 0.9$ [4]

Using the Normal distribution as an approximation to the binomial

Recap

For large values of n, probability tables for the binomial distribution may not be available and it may not be possible or practical to create tables on your calculator. For *large* values of n, the probability distribution function looks very similar to a Normal distribution curve. This works best when p is close to 0.5, otherwise the curve will not be symmetrical. However, the larger the value of n, the further away from 0.5 p is allowed to be for the approximation to be valid.

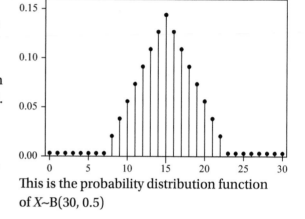

This is the probability distribution function of $X \sim B(30, 0.5)$

- $X \sim B(n, p)$ can be approximated by a Normal random variable Y, provided
  - n is large
  - p is close to 0.5
- You must use a **continuity correction** to take account of the fact that the binomial is a discrete distribution but the Normal is a continuous distribution.

If X is binomially distributed, then the table shows the continuity corrections required to approximate different probabilities using the Normal random variable Y

Original	Continuity correction
$P(X < 4)$	$P(Y < 3.5)$
$P(X \le 4)$	$P(Y < 4.5)$
$P(X \ge 4)$	$P(Y > 3.5)$
$P(X > 4)$	$P(Y > 4.5)$
$P(X = 4)$	$P(3.5 < Y < 4.5)$

- You need to calculate the mean and variance to use in the Normal approximation using
  $\mu = np$ and $\sigma^2 = np(1-p)$

$X \sim B(65, 0.45)$ can be approximated by a Normal random variable since n is large and p is close to 0.5

The mean of the Normal distribution to be used is $\mu = 65 \times 0.45 = 29.25$ and the variance is
$\sigma^2 = 65 \times 0.45 \times (1 - 0.45) = 16.0875$

So use the approximation $Y \sim N(29.25, 16.0875)$

To calculate $P(X > 30)$, you need to use a continuity correction, so calculate $P(Y > 30.5)$

Use Normal CD on your calculator with

Lower = 30.5, Upper = 999, $\sigma = \sqrt{16.0875}$, $\mu = 29.25$

This gives $P(Y > 30.5) = 0.3777$

> Remember that you need standard deviation here, not variance.

Jeremy purchased soft drinks in 28 of the 52 weeks of 2015 and wishes to use this information to predict how often he is likely to purchase soft drinks in future years.

a Give a suitable distribution to model the number of weeks, W, in which Jeremy purchases soft drinks. State any modelling assumptions you make.

b Use this distribution and a suitable approximation to estimate the probability of Jeremy purchasing soft drinks in at least half of the weeks of a year.

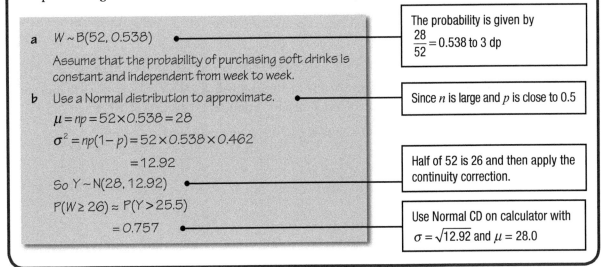

a $W \sim B(52, 0.538)$

Assume that the probability of purchasing soft drinks is constant and independent from week to week.

b Use a Normal distribution to approximate.

$\mu = np = 52 \times 0.538 = 28$

$\sigma^2 = np(1-p) = 52 \times 0.538 \times 0.462$

$= 12.92$

So $Y \sim N(28, 12.92)$

$P(W \geq 26) \approx P(Y > 25.5)$

$= 0.757$

The probability is given by
$\frac{28}{52} = 0.538$ to 3 dp

Since n is large and p is close to 0.5

Half of 52 is 26 and then apply the continuity correction.

Use Normal CD on calculator with $\sigma = \sqrt{12.92}$ and $\mu = 28.0$

The amount of skimmed milk purchased per person in a sample from London had a mean of 860 ml and a standard deviation of 230 ml. Calculate the probability that, out of 10 randomly chosen people, fewer than 2 purchase more than a litre of milk.

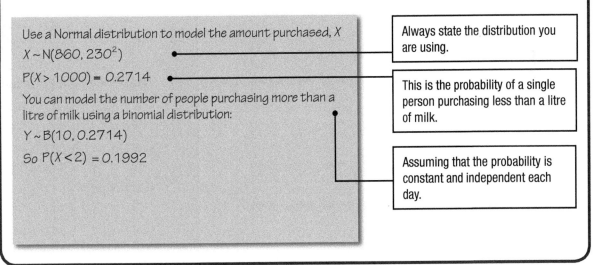

Use a Normal distribution to model the amount purchased, X

$X \sim N(860, 230^2)$

$P(X > 1000) = 0.2714$

You can model the number of people purchasing more than a litre of milk using a binomial distribution:

$Y \sim B(10, 0.2714)$

So $P(X < 2) = 0.1992$

Always state the distribution you are using.

This is the probability of a single person purchasing less than a litre of milk.

Assuming that the probability is constant and independent each day.

Exam tips

- Remember the continuity correction when approximating a binomial distribution by a Normal distribution and be careful about whether to add or subtract 0.5
- Always write down the distribution you are using for each question or question part.
- State the reasons why a Normal approximation is suitable in the context of the question.
- Make any modelling assumptions clear and relate them to the context of the question.

1   A random variable, X, is normally distributed with mean of 22 and standard deviation 3. An independent random sample of size 12 is taken from the population. Find the probability that fewer than 10 of the observations are greater than 18   [3]

2   Cans of soft drinks are labelled 'Volume 330 ml.' The machine that fills the cans is set so that the actual volume of drink dispensed has a mean of 331.1 ml with a standard deviation of 0.5 ml.

   a   Find the probability that a randomly chosen can has a volume less than 330 ml.   [3]

   b   Find the probability that, out of 12 cans checked, at least one of them is under 330 ml.   [3]

3   The random variable X has the distribution B(30, 0.65)

   a   If Y is an approximating Normal random variable and $Y \sim N(\mu, \sigma^2)$, calculate the values of μ and σ
                                                                                                                         [2]

   b   Use your results in part **a** to find approximate values of

   i   P($X = 17$)   [2]

   ii   P($X < 17$)   [2]

iii $P(17 \leq X < 22)$ [2]

_____

_____

4 The mean weight of chocolate bars bought per week by residents in the North West of England in 2012 was 85 g, with a standard deviation of 12 g.

a Assuming a Normal distribution, find the probability that a person chosen at random bought less than 80 g. [2]

_____

_____

_____

A sample of 50 people from the North West was taken and the value of the random variable Y, the number who bought less than 80 g of chocolate, was noted.

b Find the probability that Y is greater than 15. You should use an appropriate approximate distribution. [4]

_____

_____

_____

_____

c Further investigation showed that the probability found in part **b** was NOT a good estimate for the proportion of samples with Y greater than 15. Give a reason for this poor fit to the data. [1]

_____

_____

_____

5 The mean amount of bread bought per person in one week in England in 2014 was 549 g.

    a Calculate the probability that a randomly chosen person in England buys more than 600 g of bread, assuming a Normal distribution with standard deviation 42 g. [2]

A sample of 50 English residents were questioned about the amount of bread they bought the previous week.

    b State the exact distribution of Y, the number of people in the sample buying more than 600 g of bread. [1]

    c Use a Normal approximation to find the probability that more than five people in the sample bought more than 600 g of bread. [2]

6 In a large medical survey, it was found that 48% of patients had experienced breathlessness after mild exertion. A sample of 50 similar patients was taken and questioned about the symptom. Of these, 92% gave a response.

    a Use an approximate Normal distribution to find the probability that, of the patients who responded, more than 30 had experienced breathlessness. [3]

    b Using the binomial function on your calculator, calculate the probability required in part **a** and hence find the percentage error obtained when using the approximate distribution. [3]

Recap

- The **product-moment correlation coefficient**, r, gives an indication of the strength of the correlation. It is always in the range $-1 \leq r \leq 1$, so, for example, $r = 1$ indicates perfect positive correlation and $r = 0$ indicates no correlation.
- The **population correlation coefficient**, ρ, describes the actual correlation between two variables.
- The product-moment correlation coefficient, r, can be calculated from a sample and used to estimate ρ
- To test correlation, the null hypothesis is $H_0 : \rho = 0$ and the alternative hypothesis is one of
  - $H_1 : \rho \neq 0$
  - $H_1 : \rho < 0$
  - $H_1 : \rho > 0$
- In order to reject the null hypothesis, the value of r must be more extreme than the critical value.

The critical value for a 1-tailed test on a sample of size 10 at the 5% significance level is ± 0.5494

So, if the hypotheses are $H_0 : \rho = 0, H_1 : \rho > 0$, then reject the null hypothesis if $r > 0.5494$

If the hypotheses are $H_0 : \rho = 0, H_1 : \rho < 0$, then reject the null hypothesis if $r < -0.5494$

The critical value for a 2-tailed test on a sample of size 10 at the 5% significance level is ± 0.6319

So if the hypotheses are $H_0 : \rho = 0, H_1 : \rho \neq 0$, then reject the null hypothesis if $r < -0.6319$ or $r > 0.6319$

- An alternative method is to use the p-value of the statistic, r
  You need to compare the p-value with the significance level; the null hypothesis can be rejected if the p-value is less than the significance level.
- For a 2-tailed test, you need to compare the p-value to half of the significance level.

A sample of size 20 has a product-moment correlation coefficient of $r = -0.5$
The p-value for this statistic is 0.0248

To test the hypotheses $H_0 : \rho = 0, H_1 : \rho > 0$ at the 2.5% significance level:

   The null hypothesis can be rejected at the 2.5% significance level since $0.0248 < 2.5\%$

   There is no evidence of positive correlation.

To test the hypotheses $H_0 : \rho = 0, H_1 : \rho \neq 0$ at the 2.5% significance level:

   The null hypothesis cannot be rejected at the 2.5% significance level since $0.0248 > 1.25\%$
   (Notice that we are comparing to 1.25% not 2.5%, since this a 2-tailed test.)

   There is evidence of negative correlation.

Example 1

The product-moment correlation coefficient between purchases of white bread and brown bread in a sample of 20 households is found to be −0.4121

Use a hypothesis test with a 1% significance level to investigate whether there is a negative correlation between purchases of white bread and brown bread.

The critical value is −0.5155

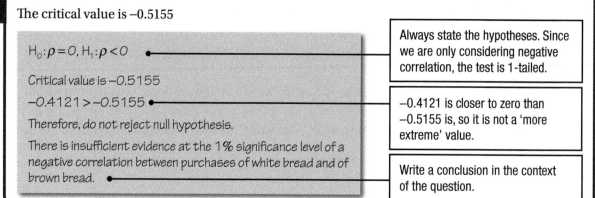

$H_0: \rho = 0$, $H_1: \rho < 0$

Critical value is −0.5155

−0.4121 > −0.5155

Therefore, do not reject null hypothesis.

There is insufficient evidence at the 1% significance level of a negative correlation between purchases of white bread and of brown bread.

Always state the hypotheses. Since we are only considering negative correlation, the test is 1-tailed.

−0.4121 is closer to zero than −0.5155 is, so it is not a 'more extreme' value.

Write a conclusion in the context of the question.

Example 2

In order to investigate whether there is any correlation between the amount of whole milk and skimmed milk purchased per person, a sample of 30 people is taken and the product-moment correlation coefficient is found to be −0.41. The p-value for this statistic is 0.024

Conduct a hypothesis test with a 5% significance level.

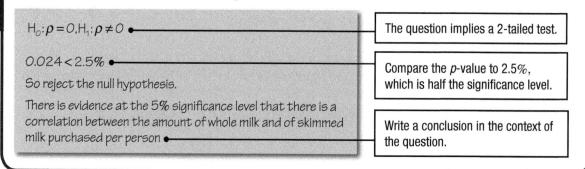

$H_0: \rho = 0$, $H_1: \rho \neq 0$

0.024 < 2.5%

So reject the null hypothesis.

There is evidence at the 5% significance level that there is a correlation between the amount of whole milk and of skimmed milk purchased per person

The question implies a 2-tailed test.

Compare the p-value to 2.5%, which is half the significance level.

Write a conclusion in the context of the question.

Exam tips

- State your hypotheses in terms of ρ; make sure you use H_0 for the null hypothesis and H_1 for the alternative hypothesis.
- Check whether the test is 1-tailed or 2-tailed.
- Write down all your steps; it must be clear how you have reached your conclusion.
- Always give your conclusions using the context of the question.

1. The correlation coefficient for two variables, X and Y, is 0.28. Test, at a 5% significance level, whether the population correlation coefficient is zero against the alternative hypothesis that it is not zero for the following sample sizes. Critical values for each test are given in brackets.

   a $n = 10$ (critical values: ±0.6319) [2]

   b $n = 20$ (critical values: ±0.4438) [2]

   c $n = 100$ (critical values: ±0.1966) [2]

2. a Two variables, R and S, have a sample correlation coefficient of 0.91 and it is believed that the population correlation coefficient between the variables is positive.

   i Write down the null and alternative hypotheses for a test of this belief. [1]

   ii Perform the test at a significance level of 2.5%, given that the sample size was 5 and the p-value corresponding to the observed sample correlation coefficient is 1.65%. [2]

3 a Given that the sample product moment correlation coefficient between variables X and Y is $r_{xy} = -0.20$, based upon a sample of size 90, test, at the 5% significance level, the hypothesis that the population correlation coefficient, ρ_{xy}, is

i Less than zero (critical value: $r = -0.1745$), [2]

_____

_____

_____

_____

ii Equal to zero (critical values: $r = \pm0.2072$). [2]

_____

_____

_____

b Explain why two different conclusions are possible from the same evidence. [2]

_____

_____

_____

_____

4 Between 2001 and 2007, white bread sales in England were in decline. Average quantities purchased for white bread and for brown/wholemeal bread in grams per person per week are given below.

Year	2001	2002	2003	2004	2005	2006	2007
White	492	456	429	383	358	320	303
Brown/wholemeal	127	128	126	175	200	201	154

4 a On the same axes, draw a scatter diagram of year versus quantity of

    i White bread,

    ii Brown/wholemeal bread purchased. [2]

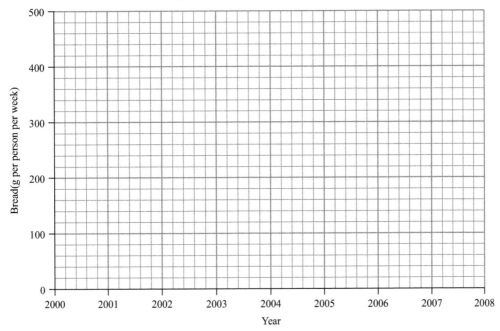

b State the sign and strength of the correlation between

    i The year and quantities of white bread bought, [1]

    ii Quantities of white bread and quantities of brown/wholemeal bread bought. [1]

The sample correlation coefficient between the year and quantities of brown/wholemeal bread bought is 0.69

c Test the hypothesis that the population from which these values came has a positive correlation. You should use a significance level of 10%. The critical value for the test is 0.5509 [2]

5 A sample is taken from a bivariate population and found to have a correlation coefficient of 0.24. You wish to test the suggestion that the population correlation coefficient between the variables is zero against the possibility that it is positive.

a Write down the null and alternative hypotheses for the test. [1]

b Given that the sample size is 34, perform the test at a significance level of 2.5%. The critical value for the test is 0.3388 [2]

c With the observed value of the sample correlation coefficient, use linear interpolation to find the approximate minimum sample size which would have resulted in the null hypothesis being rejected. (For $n = 60$, the critical value is 0.2542; for $n = 70$, the critical value is 0.2352) [3]

Recap

- If a continuous random variable, X, is Normally distributed with mean μ and variance σ^2, then write $X \sim N(\mu, \sigma^2)$
- To test the mean of a Normal distribution, the null hypothesis is $H_0 : \mu = \mu_0$ and the alternative hypothesis is one of
    - $H_1 : \mu \neq \mu_0$
    - $H_1 : \mu < \mu_0$
    - $H_1 : \mu > \mu_0$
- You need to calculate the test statistic, $z = \dfrac{\bar{x} - \mu_0}{\dfrac{\sigma}{\sqrt{n}}}$, where
    - $\bar{x}$ is the sample mean
    - σ is the standard deviation of the distribution
    - n is the sample size
    - μ_0 is the assumed mean, as stated in the null hypothesis.
- Then calculate the p-value for the test statistic using $Z \sim N(0, 1^2)$
    - If $\bar{x} > \mu$, then calculate $P(X > z)$
    - If $\bar{x} < \mu$, then calculate $P(X < z)$
- Compare this probability to the significance level and reject the null hypothesis if the probability is less than the significance level.
- If you have a 2-tailed test, then remember to halve the significance level.

If $X \sim N(\mu, 36)$, then in order to test hypotheses $H_0 : \mu = 5$, $H_1 : \mu > 5$ you first need to calculate the test statistic from a sample.

For example, if a sample of size 12 had a mean of $\bar{x} = 7$, then the test statistic is

$$z = \frac{7-5}{\dfrac{6}{\sqrt{12}}} = 1.155$$

$P(X > 1.155) = 0.1241$ (use your calculator)

This result will not be significant even at a 10% significance level.

- Alternatively, you can find the critical value using the inverse Normal function on your calculator.
- You reject the null hypothesis if the test statistic is a more extreme value than the critical value.

For the example above, the critical value at the 5% significance level is 1.6449. You can find this on your calculator using an area of 0.95, since $P(X > \text{critical value}) = 0.05$ so $P(X < \text{critical value}) = 0.95$

So if you have a test statistic greater than 1.6449, you can reject the null hypothesis.

In the case with a sample of size 12: $\dfrac{\bar{x} - 5}{\dfrac{6}{\sqrt{12}}} > 1.6449$

So values of the mean such that $\bar{x} > 7.85$ will give a significant result.

On your calculator:

Use Inverse Normal:

Calculator

Area : 0.95
$\sigma : 1$
$\mu : 0$

xInv = 1.6449

Example 1

The distribution of cheese purchased per person in England is thought to follow a Normal distribution with mean 113 g. The standard deviation is assumed to be 88 g. A sample of 50 people from the South West is taken and the average amount of cheese purchased per person is 139 g.

Test, at the 5% significance level, whether the mean is different in the South West compared with the whole of England.

| | Always state your hypotheses. |

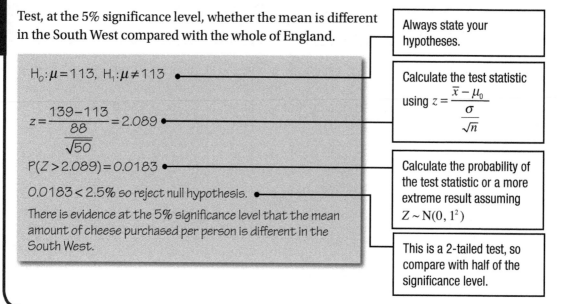

$H_0: \mu = 113, \ H_1: \mu \neq 113$

$z = \dfrac{139 - 113}{\dfrac{88}{\sqrt{50}}} = 2.089$

$P(Z > 2.089) = 0.0183$

$0.0183 < 2.5\%$ so reject null hypothesis.

There is evidence at the 5% significance level that the mean amount of cheese purchased per person is different in the South West.

Calculate the test statistic using $z = \dfrac{\bar{x} - \mu_0}{\dfrac{\sigma}{\sqrt{n}}}$

Calculate the probability of the test statistic or a more extreme result assuming $Z \sim N(0, 1^2)$

This is a 2-tailed test, so compare with half of the significance level.

Example 2

The amount of salt in a large portion of chips from a fast-food restaurant is Normally distributed with variance $0.4\,g^2$ and intended mean 1 g.

Complaints have been received that the restaurant is using too much salt, so a sample of 25 portions is taken and a hypothesis test is conducted to see if the complaints are justified.

a State the null and alternative hypotheses.

b Work out the range of values of $\bar{x}$ that would lead to the assumption that the mean is higher than 1 g at the 1% significance level.

a $H_0: \mu = 1, \ H_1: \mu > 1$

b $P(Z > 2.3263) = 0.01$

So $z = \dfrac{\bar{x} - 1}{\dfrac{\sqrt{0.4}}{\sqrt{25}}} > 2.3263$

$\Rightarrow \bar{x} > 1.29\,g$

Using inverse Normal on calculator with area = 0.99

Use $z = \dfrac{\bar{x} - \mu_0}{\dfrac{\sigma}{\sqrt{n}}}$ to form an inequality in $\bar{x}$

Exam tips

- Give your hypotheses in terms of μ; make sure you use H_0 for the null hypothesis and H_1 for the alternative hypothesis.
- Check whether the test is 1-tailed or 2-tailed.
- Show your method for calculating the test statistic.
- Write down either
  - the critical value or
  - the probability of obtaining the test statistic or a more extreme value.

  It must be clear how you have reached your conclusion.
- Always give your conclusions in the context of the question.

1 A Normally distributed population has a variance of 12, but the mean is unknown. A sample of size 50 taken from this population has a mean value of 14.1. You wish to test whether the population mean is greater than 13

Let X be a random variable for the population values.

  a State the null and alternative hypotheses for the test. [1]

  b Give the distribution of the sample mean, $\bar{X}$, assuming the null hypothesis to be true. [2]

  c Perform the test at a significance level of 2.5%, stating clearly your conclusion. [3]

2 The mean amount of semi-skimmed milk purchased per person per week in 2008 was 999 ml, with a standard deviation of 63 ml. In 2013, a sample of 45 individuals was taken and the amount purchased by each individual was recorded. The mean of these values was 1010 ml.

You wish to test the hypothesis that the quantity of semi-skimmed milk had changed during these five years.

Let the population values in 2013 be described by the random variable X and let the population mean be μ

  a Give, in terms of μ, the distribution of X. You should assume that the population variance remains unchanged between 2008 and 2013 [1]

2 b State the null and alternative hypotheses in the test and, assuming that the null hypothesis is true, give the distribution of $\bar{X}$, a random variable for the sample mean. [2]

c By calculating a suitable p-value, perform the test at a 10% significance level and state clearly your conclusion to the test. [3]

3 A sample of size 10 is taken from a Normal population and gives the following values.

212.4,   201.7,   227.5,   210.6,   203.1,   217.5,   230.9,   225.2,   209.3,   227.3

a Calculate the sample mean. [1]

b By finding the critical values of the sample mean, test whether the population mean is 208. Assume that the standard deviation of the population is 12. You should use a significance level of 5%. [3]

4 An investigation comparing the amount of mineral water bought by Londoners and residents of the rest of the South East found that the mean quantity of water bought in London in 2014 was 347 ml per person per week. In the South East, a sample of 55 residents bought an average of 335 ml of water.

Test, at a significance level of 10%, whether the two regions differ in the quantity of water bought, stating any assumptions you make.

You should assume that the standard deviation of the quantity of water bought is equal to 42 ml for both regions.

[6]

5 Analysis of several texts shows that the mean number of words in sentences written by Charles Dickens is 17.66. A linguist researching the origins of a recently discovered fragment of a novel counts the number of words in each sentence of the fragment. The results are given in the following table.

Number of words per sentence	Number of sentences
9–11	5
12–14	8
15–17	12
18–20	9
21–23	5
24–26	1

5 a Find the mean and the variance of the number of words for the fragment's sentences. [2]

The researcher wishes to use these results to test whether the fragment of text could be by Charles Dickens.

b State the null and alternative hypotheses for the test. [1]

c Perform the test at a 5% significance level, stating any assumptions you make. Assume that the population variance of the sentence lengths is equal to the sample variance. [3]

Track your progress

Use these checklists to track your confidence level in each topic of A Level Statistics.

Ch	Objective	MyMaths	InvisiPen	Not yet	Almost	Yes!
9 Collecting, representing and interpreting data	Distinguish a population and its parameters from a sample and its statistics.	2275	09S1A	☐	☐	☐
	Identify and name sampling methods.	2275	09S1A	☐	☐	☐
	Highlight sources of bias in a sampling method.	2275	–	☐	☐	☐
	Read continuous data given in box-and-whisker plots, histograms and cumulative frequency diagrams.	2276–2278	09S3A	☐	☐	☐
	Plot scatter diagrams and use them to identify types and strength of correlation.	2283	09S4B	☐	☐	☐
	Use scatter diagrams and rules using quantities to identify outliers.	2283	09S4B	☐	☐	☐
	Summarise raw data using appropriate measures of location and spread.	2279–2282	09S2B	☐	☐	☐
10 Probability and discrete random variables	Use the vocabulary of probability theory, including the following terms: random experiment, sample space, independent events and mutually exclusive events.	2093	10S1B	☐	☐	☐
	Solve the problems involving mutually exclusive and independent events using the addition and multiplication rules.	2093, 2094	10S1B	☐	☐	☐
	Use a probability function or a given context to find the probability distribution and probabilities for particular events.	2114	–	☐	☐	☐
	Recognise and solve problems relating to experiments which can be modelled by the binomial distribution.	2110, 2111	10S2A	☐	☐	☐
11 Hypothesis testing 1	Understand the terms null hypothesis and alternative hypothesis.	2115	11S1A, 11S2B	☐	☐	☐
	Understand the terms critical value, critical region and significance level.	2115	11S1A	☐	☐	☐
	Calculate the critical region.	2115	11S1A	☐	☐	☐
	Calculate the p-value.	2115	–	☐	☐	☐
	Decide whether to reject or accept the null hypothesis.	2115	11S1A	☐	☐	☐
	Make a conclusion based on whether you reject or accept the null hypothesis.	2115	11S2B	☐	☐	☐

Track your progress

Use these checklists to track your confidence level in each topic of A Level Statistics topics.

Ch	Objective	MyMaths	InvisiPen	No	Almost	Yes!
20 Probability and continuous random variables	Calculate conditional probabilities from data given in different forms.	2092, 2095	20S1A	☐	☐	☐
	Apply binomial and Normal probability models in different circumstances.	2113, 2120, 2121, 2292	20S2B, 20S3A	☐	☐	☐
	Use data to assess the validity of probability models.	2286	–	☐	☐	☐
	Solve problems involving both binomial and Normal distributions.	2286	20S4B	☐	☐	☐
21 Hypothesis testing 2	State null and alternative hypotheses when testing for correlation.	2287	21S1A	☐	☐	☐
	Compare a given PMCC to a critical value or its p-value to the significance level, and use this comparison to decide whether to accept or reject the null hypothesis.	2287	21S1A	☐	☐	☐
	Decide what the conclusion means in context about the correlation.	2287	21S1A	☐	☐	☐
	State null and alternative hypotheses when testing the mean of a Normal distribution.	2288	21S2B	☐	☐	☐
	Calculate the test statistic, compare it to a critical value or compare its p-value to the significance level, and use this comparison to decide whether to accept or reject the null hypothesis.	2288	21S2B	☐	☐	☐
	Decide what the conclusion means in context about the mean of the distribution.	2288	21S2B	☐	☐	☐

Formulae you will be given

Probability and Statistics

Probability

$P(A \cup B) = P(A) + P(B) - P(A \cap B)$

$P(A \cap B) = P(A) \times P(B|A)$

Standard deviation

$$\sqrt{\frac{\Sigma(x - \bar{x})^2}{n}} = \sqrt{\frac{\Sigma x^2}{n} - \bar{x}^2}$$

Discrete distributions

Distribution of X	$P(X = x)$	Mean	Variance
Binomial B(n, p)	$\binom{n}{x} p^x (1-p)^{n-x}$	np	$np(1-p)$

Sampling distributions

For a random sample of n observations from $N(\mu, \sigma^2)$:

$$\frac{\bar{X} - \mu}{\frac{\sigma}{\sqrt{n}}} \sim N(0, 1)$$

Formulae you need to learn

The mean of a set of data: $\bar{x} = \dfrac{\Sigma x}{n} = \dfrac{\Sigma fx}{\Sigma f}$

The standard Normal variable: $Z = \dfrac{X - \mu}{\sigma}$ where $X \sim N(\mu, \sigma^2)$

Section 9.1

1 a All students in year group.
   b Stratified random sampling.
   c Number of boys = 120 − 80 = 40
      10% of 40 = 4; 10% of 80 = 8
      4 boys, 8 girls.
2 a Cheap and convenient yet still unbiased.
   b Method in **a** would be biased, applying to only one temperature. Systematic sampling throughout the day would ensure sampling at all temperatures.
3 a Stratified random sampling and cluster sampling. Stratifying according to region allows for differences in meat consumption between regions; using households as clusters means that a list of all adults in the UK is not required.
   b The sample might have disproportionate number of households from certain regions.
4 a Bias towards country rather than city dwellers. Proportion of males to females may not reflect nationwide proportions.
   b Use a variety of country and urban locations for sampling. Stratify according to gender.
5 Obtain a list of all students in the school – the sampling frame. Arrange into two lists of boys and girls, each grouped in classes. Choose a number at random from 1–10; take that numbered boy and every subsequent 10th boy. Repeat for the girls.
6 a Fruit eating is seasonal. Incorrect conclusions could be drawn, for example, if strawberry consumption in August 2010 were compared with that in February 2011
   b It can't be assumed that patterns of fruit consumption are similar for different countries. Small samples are unpredictable and therefore it is likely that proportions for the different countries in the sample will differ from those throughout the UK generally. Stratification by country guarantees the correct proportions.
7 a Population size is 2597 + 7052 + 5284 + 4533 + 5602 = 25068
      Number of claimants in the North East = 3% of 2597 = 78
      Number of claimants in the North East in the sample
      $\left(\dfrac{78}{25068}\right) \times 1000 = 3.1115... = 3$ claimants
      Number of non-claimants in the North East = 2597 − 78 = 2519
      Number of non-claimants in the North East in the sample
      $\left(\dfrac{2519}{25068}\right) \times 1000 = 100.4866... = 100$ claimants
   b 974 litres (to nearest litre)

Section 9.2

1 a Median: 5.5
      IQR: 5.8
   b Median: 2
      IQR: 2
2 a $\bar{x} = 13.75 \quad s^2 = 12.02$ (2 dp)
   b $\bar{x} = -0.12 \quad s^2 = 0.99$ (2 dp)
3 67.73 (2 dp)
4 a $\bar{x} = 251.86$ (2 dp)
      $s^2 = 1358.41$ (2 dp)
      $s = 38.86$ (2 dp)
   b 178.14, 325.57 (2 dp)
      163 is an outlier.
5 a $\bar{x} = 90.40$ (2 dp)
      $s^2 = 35.12$ (2 dp)
   b 0.58 (2 dp)
6 a Town A: $\sum x = 2279.7$
      $x^2 = 309485.17$

      Town B: $\sum x = 1041.6$
      $\sum x^2 = 137231.52$

   b $\bar{x} = 132.85$ (2 dp)
      $s^2 = 219.01$ (2 dp)
   c Mean of the means $= \dfrac{(134.1 + 130.2)}{2} = 132.15 \neq 132.85$
      Different sample sizes
   d $\bar{x}_w = \dfrac{n_1 \bar{x}_1 + n_2 \bar{x}_2}{n_1 + n_2} = \dfrac{17 \times 134.1 + 8 \times 130.2}{17 + 8} = 132.85$ (2 dp)

Section 9.3

1 a

	Median	Range	Interquartile range
Regular	1510	880	530
Low calorie	510	1010	550

   b A lower amount of low-calorie drinks were purchased. The spread of the amount of low-calorie purchase is greater than for regular drinks (measured by both the the range and interquartile range).

2 a

Interval, g	$0 \le g$ < 20	$20 \le g$ < 40	$40 \le g$ < 50	$50 \le g$ < 55	$55 \le g$ < 60	$60 \le g$ < 70	$70 \le g$ ≤ 90
Frequency	4	8	7	6	4	3	2

   (Bar drawn between 40 and 50 with a height of 7)
   b 47.86 grams (2 dp)
   c $\dfrac{15}{34}$
3 a i 2.8
      ii 4.3
      iii 6
   b 4 days
4 a

Blood sugar, b (mmol/litre)	Before	After
$3 \le b < 4$	1	0
$4 \le b < 5$	6	2
$5 \le b < 6$	9	8
$6 \le b < 7$	7	7
$7 \le b < 8$	0	4
$8 \le b < 9$	0	2

   b Before:
      median 5.5
      lower quartile 4.85; upper quartile 6.15
      After:
      median 6.2
      lower quartile 5.5; upper quartile 7.05
   c 'After' readings are on average larger (as measured by median) and with greater spread (larger interquartile range) (1.55 compared to 1.3).

Section 9.4

1 A: 4   B: 3   C: 1   D: 2
2 a

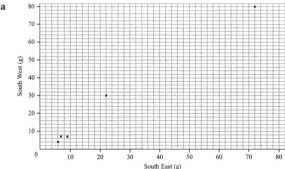

   Strong positive correlation.

b 0.8

c Spurious correlation occurs when two variables show a relationship but there is no causal connection between them. x and y have spurious correlation because the correlation coefficient is non-zero but levels of SE consumption don't cause equivalent values in SW.

3 a

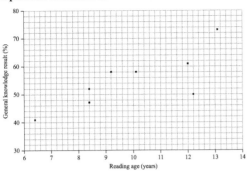

b Moderate positive correlation.

c 18 years is outside the given range of reading ages.

d Accept 56% - 60%

4 a i Positive correlation
  ii Negative correlation
  iii Positive correlation
  iv Negative correlation

b N_1 and N_C. C occurs if and only if 1 occurs.

5 a

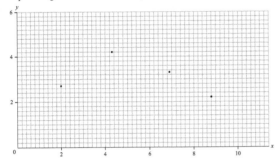

b i A 5.4
  ii D 4.4

Section 10.1

1 a i Independent. Two events where the outcome of one has no effect on the probability of the other occurring.
  ii Mutually exclusive. It is impossible for both events to occur simultaneously.
  iii Mutually exclusive. It is impossible for both events to occur simultaneously.

b One of the events has to occur when the experiment is performed.

2 a Define H_1: head on first throw; $T_1$1: tail on first throw, 1 on dice, etc.
  H_1H_2 H_1T_2 T_11 T_12 T_13 T_14 T_15 T_16

b The outcomes are not all equally likely.

3 a $\dfrac{5}{36}$

b $\dfrac{5}{432}$

4 a $P(A \cap B) = P(A) \times P(B)$
  or $P(A) = P(A \mid B)$ or $P(B) = P(B \mid A)$

b i 0.27
  ii 0.54
  iii 0.61
  iv From parts **a** and **b**, P(sodium intake less than 2.9 grams per day) × P(less than 50 years old) = 0.54 × 0.61 = 0.3294
  From table, P(sodium intake less than 2.9 grams per day and less than 50 years old) = 0.34
  P(sodium intake less than 2.9 grams per day and less than 50 years old) ≠ P(sodium intake less than 2.9 grams per day) × P(less than 50 years old)

Therefore, a person's sodium intake is not independent of the person's age.

5 a $a = \dfrac{1}{6}$
  $P(X < 4) = \dfrac{1}{2}$

b Assume that X values are independent of each other.
  $\dfrac{3}{8}$

6 a $\dfrac{1}{8}$

b $\dfrac{3}{8}$

Section 10.2

1 a 0.296
  b 0.552
  c 0.953

2

x	0	1	2	3	4	5	6
P($X = x$)	0.016	0.094	0.234	0.313	0.234	0.094	0.016

3 a 0.25
  b 0.22
  c 0.92

4 a $X \sim B\left(7, \dfrac{1}{6}\right)$

b i 0.234
  ii 0.096
  iii 0.904

5 X = the number which are either red or white. $X \sim B(8, 0.7)$
  $P(X = 6) = 0.296$ (3 dp)

6 0.536 (3 dp)

7 0.118 (3 dp)
  Assume sample is made up of 15 independent English residents.

8 a 0.1
  b 0.176
  c Let required number of adults be n. $X \sim B(n, 0.1)$
  It is required that $P(X \geq 5) \geq 0.9$
  For $n = 78$, $P(X \geq 5) = 0.901$; for $n = 77$, $P(X \geq 5) = 0.895$
  Therefore 78 adults should be questioned.

9 a Median is $8.5 + \dfrac{4}{6} \times 1 = 9\dfrac{1}{6}$
  b $\dfrac{1}{2}$
  c 0.313 (3 dp)
  d Results only approximate since population size (31) not very large compared with sample size (4)

10 a 0.549 (3 dp)
  b 154 children

Section 11.1

1 $X > 23$

2 a H_0: $p = 0.64$ H_1: $p \neq 0.64$ where p is the proportion in the population who think that local crime is increasing.
  b $X \sim B(45, 0.64)$

3 a Use an independent random sample.
  b H_0: $p = 0.22$ H_1: $p \neq 0.22$ where p is the proportion in the population who are trading down.
  c $X \sim B(40, 0.22)$
  d 14 is the critical value and therefore within the critical region, so the null hypothesis should be rejected (22% is not the correct proportion of households trading down).

4 a The critical region is the set of extreme values in the distribution defined in the null hypothesis that have a total probability equal to or less than the significance level of the test, and which are more likely under the alternative hypothesis. These values would lead to the null hypothesis being rejected because they are unlikely given that the null hypothesis is true.
  b i H_0: $p = 0.5$ H_1: $p < 0.5$ where p is the proportion in the population in 2015 who bought more than 280 g of white bread per person per week.
  $X \sim B(16, 0.5)$
  ii 4 is within the critical region, therefore reject the null hypothesis: sales of white bread have indeed fallen.

5 a The experiment must consist of a fixed number of independent and identical trials where the random variable is the number times a particular outcome occurs.

b $P(X = x) = \binom{n}{x} p^x (1-p)^{n-x}; x = 0, 1, ..., n$

6 a Members of the sample should be independent of each other, and chosen randomly from the population of English residents. This is because certain individuals may be affected by being from a similar area, or related to (etc.) other members of the sample.

b $H_0: p = 0.24$, $H_1: p < 0.24$ where p is the proportion in the population who purchase more than 380 ml
$X \sim B(28, 0.24)$

c 3 is within the critical region. Reject the null hypothesis: evidence suggests that the amount purchased of this type of milk has decreased.

Section 11.2

1 a $H_0: p = 0.72$ $H_1: p < 0.72$
b 0, 1 because $P(X \leq 1) = 0.0079$ (< 0.05) and $P(X \leq 2) = 0.056$ (> 0.05)

2 a Let X be the number in the sample with higher than 70% of calorie intake derived from fatty acids.
$X \sim B(16, p)$
$H_0: p = 0.7$ $H_1: p > 0.7$
Under H_0, $P(X \geq 14) = 0.099$ (2 sf) (< 0.1)
Reject H_0; there is evidence to accept the claim that the proportion has increased.

b Conclusion should be treated cautiously because 10% is a large significance level and the probability of the observed value under the null hypothesis is very close to 0.1. That is, the probability of rejecting the null hypothesis when it is true (a type 1 error) is relatively high.

3 a $H_0: p = 0.6$ $H_1: p \neq 0.6$
Where p = the proportion of people who agree their butter purchase has increased since the last year.

b Define X, the number in the sample who agreed with the proposition.
$X \sim B(25, p)$
Under H_0, $P(X \geq 20) = 0.029$ (2 sf) (> 0.025)
At a significance level of 5%, there is no reason to reject the null hypothesis. Insufficient evidence that the proportion agreeing with the proposition has changed.

4 Let X be the number in the sample who bought less than 1206 ml per week.
$X \sim B(35, p)$
$H_0: p = 0.5$ $H_1: p \neq 0.5$
Under H_0, $P(X \leq 12) = 0.045$ (2 sf) (> 0.025)
Do not reject H_0; There is insufficient evidence to suggest a change in the amount of soft drinks purchased.

5 a There is a fixed sample size and it seems reasonable to assume that the probability of any one egg hatching is the same as, and independent of, any other.

b

x	39	40	41	≥ 42
$P(X = x)$	0.036	0.024	0.016	0.019

c Let X be the number in the sample that hatch.
$X \sim B(64, p)$
$H_0: p = 0.52$ $H_1: p > 0.52$
Under H_0, $P(X \geq 39) = 0.095$ (2 sf) (> 0.05)
No reason to reject H_0; there is no evidence to suggest an increase in the proportion hatching.

6 a It is reasonable to assume that heights are normally distributed, and for a normal distribution, mean = median. Therefore, proportion above mean is 0.5

b $X \sim B(40, p)$
$H_0: p = 0.5$ $H_1: p > 0.5$
Under H_0, $P(X \geq 27) = 0.019$ (2 sf) (< 0.02)
Reason to reject H_0; there is evidence to suggest that Irish males were taller than English males.

Section 20.1

1 a Let W_1 be the event 'white from box 1',...

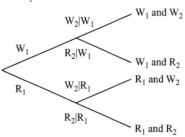

b $\dfrac{5}{8}$

2 a i $\dfrac{11}{25}$

ii $\dfrac{7}{11}$

b P(BMI less than 25 and less than the median amount of sugar)
$= \dfrac{8}{50} = 0.16$ (A)
P(BMI less than 25) $= \dfrac{11}{50} = 0.22$
P(less than the median amount of sugar) $= \dfrac{22}{50} = 0.44$
P(BMI less than 25) × P(less than the median amount of sugar)
$= 0.22 \times 0.44 = 0.0968$ (B)
Answers A and B very different suggesting that the events are not independent.

3 a

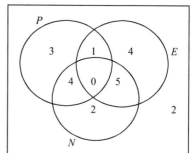

b i 0
ii $\dfrac{2}{21}$
iii $\dfrac{5}{11}$

4 a 0.67
b 0.067
c 0.828
d 0.081 (2 sf)

5 a 0.0992
b 0.21 (2 sf)

Section 20.2

1 a

Month	Jan	Feb	Mar	Apr	May	June
Receipts (£000s)	14.7	14.7	14.7	14.7	14.7	14.7

b This belief seems unlikely; actual sales are higher than expected at the early and late months, lower for the central months.

2 a

Year	2005	2006	2007	2008	2009	2010	2011	2012	2013	2014
Amount of bread (g)	629.1	629.1	629.1	629.1	629.1	629.1	629.1	629.1	629.1	629.1

b No, the data does not suggest that bread purchasing has remained constant over 10 years.
Expected frequencies at either end of the period differ significantly from the observed values.
Bread purchases for every year from 2006 to 2016 show a decrease on the previous year. Even if all differences between observed and expected frequencies were small, this is extremely unlikely under the hypothesis that sales are not changing.

3 a

Colour	Red	Blue	Black
Probability	$\dfrac{1}{8}$	$\dfrac{1}{4}$	$\dfrac{5}{8}$

b Theory suggests expected frequencies 7.5, 15, 37.5. Results appear to support the theory; maximum discrepancy between observed and expected frequency is 20% of expected frequency.

4 a Trials are not identical; for example, men and women in a sample might follow different statistics in relation to the outcomes. Trials are not independent in relation to the outcomes; for example, the sample may consist of members of one family.

b

X	0	1	2	3	4	5	6
$P(X=x)$	0.047	0.187	0.311	0.276	0.138	0.037	0.004

c Expected frequencies based upon belief:

x	0	1	2	3	4	5	6
Frequency	4.7	18.7	31.1	27.6	13.8	3.7	0.4

Very poor agreement between actual and theoretical frequencies suggest belief is not true. Observed frequencies are low for small x and high for large x suggesting that the true value of p is greater than 0.4

5 a $X \sim B(50, 0.5)$. Assume that 350 ml is the median quantity of mineral water bought.

b $50 \times 0.5 = 25$

c $P(X \geq 33) = 0.016$ (2 sf)

d Sales are very likely to have increased. The observed value of 33 people who have bought over the previous median value is much larger than the previous expected value of 25. Also, the probability of being this far away from 25 is very small assuming no increase. In a hypothesis test (with $H_0: p = 0.5$, $H_1: p > 0.5$), the null hypothesis is likely to be rejected suggesting a further increase in sales of mineral water.

Section 20.3

1 a 0.8849

b 0.0864

c 0.5372

2 a 0.7881

b 0.6730 or 0.6731

c 0.4967 or 0.4968

3 5.531 (3 dp)

4 a 0.58 (2 dp)

b 0.90 (2 dp)

5 a $X \sim N(\mu, \sigma^2)$
$P(\mu - 2\sigma < X < \mu + 2\sigma) = P(-2 < Z < 2)$
$= 2\,P(0 < Z < 2)$
$= 2\{P(Z < 2) - P(Z < 0)\}$
$= 2(0.9772 - 0.5) = 0.9545$ (approx 95%)

b i Mean 77.167 standard deviation 35.016

ii Mean ± 2 standard deviations = (7.1342, 147.199)

Number less than 7.1342 $= \left(\dfrac{7.1342}{30}\right) \times 6 = 1.4268$

Number less than 147.199 $= 52 + \dfrac{27.199}{30} \times 8 = 59.253$

Proportion within the interval (7.135, 147.199) is

$\dfrac{(59.2530 - 1.4268)}{60} = 0.96$ (2 dp)

c The answers to parts **a** and **b ii** are very close to each other, suggesting an underlying Normal distribution. X takes values from approximately 0 to 150 and the mean value is close to the centre of this interval, consistent with a Normal distribution. Finally, quantities such as this whose values depend upon many factors are often Normally distributed. It is very likely that this variable is approximately Normally distributed.

6 a i $\bar{x} = 39.05$; $s = 12.4741$ (4 dp)

ii 1.6277, 76.4723

b The variable can be considered as continuous, all values within 3 standard deviations of the mean and some symmetry (smallest value – mean – largest value) shown.

c 0.19 (2 dp)

7 a $\mu_X = 8.43$ (2 dp)

b $\mu_Y = 9.92, \sigma_Y = 3.96$ (2 dp)

Section 20.4

1 0.089 (2 sf)

2 a 0.014 (2 sf)

b 0.15 (2 dp)

3 a $\mu = 19.5$; $\sigma = 2.612$ (3 dp)

b i 0.10 (2 dp)

ii 0.13 (2 dp)

iii 0.65 (2 dp)

4 a 0.3384

b 0.66 (2 dp)

c Assumption that the amount of chocolate bought was normally distributed was not justified.

5 a 0.1123

b $Y \sim B(50, 0.1123)$

c 0.52 (2 dp)

6 a 0.0065 (2 sf)

b 4.84 (2 dp)

Section 21.1

1 a $H_0: \rho = 0$. $H_1: \rho \neq 0$
$-0.6319 < 0.28 < 0.6319$. No reason to reject H_0

b $H_0: \rho = 0$. $H_1: \rho \neq 0$
$-0.4438 < 0.28 < 0.4438$. No reason to reject H_0

c $H_0: \rho = 0$. $H_1: \rho \neq 0$
$0.28 > 0.1966$; Reject H_0

2 a i $H_0: \rho = 0$. $H_1: \rho > 0$

ii $1.65 < 2.5\%$; reject H_0. Evidence suggests R and S have positive correlation.

3 a i $H_0: \rho = 0$ $H_1: \rho < 0$
$-0.20 < -0.1745$. Reject H_0; evidence suggests that the population has a negative correlation.

ii $H_0: \rho = 0$ $H_1: \rho \neq 0$
$-0.20 > -0.2072$. No reason to reject H_0; evidence suggests that the population has zero correlation.

b Conclusions depend not only upon the evidence but on the questions being asked. In this case, different questions are asked in parts **i** (is ρ zero or less than zero?) and **ii** (is ρ zero or not equal to zero?). Different critical values due to one- and two-tailed tests.

4 a

b i Strong negative correlation.

ii Moderate negative correlation.

c $0.69 > 0.5509$. Reject H_0; evidence suggests that the quantity of brown/wholemeal bread bought was increasing over this period.

5 a $H_0: \rho = 0$ $H_1: \rho > 0$

b $0.24 < 0.3388 \rightarrow$ No reason to reject the null hypothesis. Insufficient evidence to suggest positive correlation.

c For $n = 60 \rightarrow$ do not reject H_0.
For $n = 70$, critical value is $0.2352 \rightarrow$ reject H_0.
Sample size corresponding to 2.5% critical region is

$60 + \dfrac{(0.2542 - 0.24)}{(0.2542 - 0.2352)} \times 10 = 67.4736$

For a significant result (null hypothesis rejected), sample size should be 68

Section 21.2

1 a $X \sim N(\mu, 12)$ where μ is the population mean.
$H_0: \mu = 13$; $H_1: \mu > 13$

b $\bar{X} \sim N\left(13, \dfrac{12}{50}\right)$

c $Z = \dfrac{\bar{X} - \mu}{\frac{\sigma}{\sqrt{n}}} = \dfrac{14.1 - 13}{\sqrt{\frac{12}{50}}} = 2.25$ (2 dp)

Under H_0, $Z \sim N(0, 1)$
$2.25 > 1.96$. Reject H_0; evidence supports mean is more than 13

2 a $X \sim N(\mu, 63^2)$

b $H_0: \mu = 999$; $H_1: \mu \neq 999$ $\bar{X} \sim N\left(\mu, \dfrac{63^2}{45}\right)$

c Under the null hypothesis $\bar{X} \sim N\left(999, \dfrac{63^2}{45}\right)$

$p = P(\bar{X} > 1010) = 0.12$ (2 dp)
$0.12 > 0.05$ (10% two-tailed test).
No reason to reject H_0; evidence supports the hypothesis that
the quantity of semi-skimmed milk purchased has not changed.

3 a $\bar{x} = 216.55$

b Let X be a random variable for the population values.
$\bar{x} = 216.55$
$X \sim N(\mu, 12^2)$ where μ is the population mean. $\bar{X} \sim N\left(\mu, \dfrac{12^2}{10}\right)$
$H_0: \mu = 208$; $H_1: \mu \neq 208$
Under H_0, critical values are $208 \pm 1.96 \times \dfrac{12}{\sqrt{10}}$
$= 200.56, 215.44$ (2 dp)
$216.55 > 215.44 \rightarrow$ Reject H_0; evidence suggests population
mean is not 208

4 Let X be a random variable for the quantity of water in millilitres
bought by a randomly chosen person in the South East.
Assume that X has a normal distribution.
$X \sim N(\mu, 42^2)$ where μ is the population mean.

$\bar{X} \sim N\left(\mu, \dfrac{42^2}{55}\right)$

$H_0: \mu = 347$; $H_1: \mu \neq 347$

$Z = \dfrac{\bar{X} - \mu}{\frac{\sigma}{\sqrt{n}}} = \dfrac{335 - 347}{\frac{42}{\sqrt{55}}} = -2.12$ (2 dp)

Under H_0, $Z \sim N(0, 1)$
$-2.12 < -1.645$. Reject H_0; evidence supports difference in
quantities of water bought.

5 a $\bar{x} = 16.3$
$s^2 = 14.76$

b Let X be a random variable for the number of words in a
randomly chosen sentence and let $E(X) = \mu$
$H_0: \mu = 17.66$; $H_1: \mu \neq 17.66$

c Assume that X has a Normal distribution.
$N(\mu, 14.76)$ where μ is the population mean.

$\bar{X} \sim N\left(\mu, \dfrac{14.76}{40}\right)$

Under H_0, $\bar{X} \sim N\left(17.66, \dfrac{14.76}{40}\right)$

Under H_0, critical values $17.66 \pm 1.96 \times \sqrt{\dfrac{14.76}{40}} = 16.47$,

18.85 (2 dp)
$16.3 < 16.47$. Reject H_0; evidence supports suggestion that
Charles Dickens is not the author of the fragment.